George R. Morehouse, Silas Weir Mitchell

Researches Upon the Anatomy and Physiology of

Respiration in the Chelonia

.

George R. Morehouse, Silas Weir Mitchell

Researches Upon the Anatomy and Physiology of Respiration in the Chelonia

ISBN/EAN: 9783337184094

Printed in Europe, USA, Canada, Australia, Japan

Cover: Foto ©berggeist007 / pixelio.de

More available books at **www.hansebooks.com**

RESEARCHES

UPON THE

ANATOMY AND PHYSIOLOGY OF RESPIRATION

IN THE

CHELONIA.

BY

S. WEIR MITCHELL, M. D. AND GEORGE R. MOREHOUSE, M. D.

[ACCEPTED FOR PUBLICATION, MARCH, 1863.]

COMMISSION

PREFACE.

WITH certain slight exceptions, which we have pointed out in the text, the following essay is in the strictest possible sense the joint production of its two authors, who are equally responsible for all of its statements.

The woodcuts owe much of their accuracy to the skill of the engraver, Mr. Wilhelm, to whose experience as an anatomical draughtsman the authors are under obligation.

They entertain the wish that the novel views of the present paper may induce comparative anatomists and physiologists to examine afresh the respiratory mechanism of other reptiles, and also of birds—a labor in which they indulge the hope of sharing.

S. WEIR MITCHELL,
No. 1226 Walnut Street, Philadelphia.

GEO. R. MOREHOUSE,
No. 227 South Ninth Street, Philadelphia.

TABLE OF CONTENTS.

CHAPTER II.

PHYSIOLOGY OF THE RESPIRATORY APPARATUS OF CHELONIA.

LIST OF WOOD-CUTS.

RESEARCHES

UPON THE

ANATOMY AND PHYSIOLOGY OF RESPIRATION IN THE CHELONIA.

CHAPTER I.

In the whole animal series there is scarcely a creature that would be less likely to suggest itself as a field for discovery than the Turtle. Its temptingly curious form, its world-wide distribution, its limited means of escape and of defence, would seem to combine to render it an easy and early object of investigation to the naturalist. And yet the history of Chelonians is full of discordant observations; functions have been misinterpreted, and even important parts of structure have been asserted to exist by some, and again denied by others, until at the present day the uncertain record has forced opinion into error, and permitted the conduct of one of the most important processes of life, that of respiration, to remain misunderstood, and the means of its accomplishment neglected and in part unknown. The view now entertained by the leading authorities upon the subject, that Turtles inspire by an act of deglutition, as do the frogs, has prevailed from the first, and doubtless arose from the panting movements of the under part of the throat, common to both orders, and among turtles, especially observable in marine species. It will be the object of this paper to show that this view is incorrect, that turtles do not swallow the air in breathing, but that their respiratory act is effected by inspiratory and expiratory muscles situated within the trunk.

The solid thorax clearly indicates that Chelonians do not enjoy the perfect respiratory mechanism of the highest vertebrates. The ordinary tranquil respiration of mammals, when the ribs are at rest and the cavity of the thorax is enlarged by the descent of the diaphragm alone, is, however, very strikingly analogous to that of turtles, in which the cavity of the shield is enlarged by the contraction of the muscles of the flanks.

In tracing the anatomical history of the organs of respiration in Chelonians, the earliest work to which we have had access is a "Dissertation on the Respiration of the Tortoise," by Robert Townson, LL. D., written at Göttingen, May, 1795; and as we find in it a brief review of all that was known previously upon the subject, we have taken the privilege of embodying this rare and interesting paper in the present sketch. This we do more cheerfully as an act of justice to the author; for, having conducted our inquiry with a full knowledge of the opinions of modern

1

authorities, we were surprised, on afterwards learning the singularly truthful views of Townson, to find they had fallen unappreciated, and that, in many instances, they had not even been honored by a notice, or, when so noticed, had been mentioned only to be condemned.

PHYSIOLOGICAL OBSERVATIONS ON THE AMPHIBIA. *Dissertation the Third, on the Respiration of the Tortoise.* ROBERT TOWNSON.

The first inspection of the structure of the animals I have lately treated of, the Ranæ and Salamandræ of Linnæus, will show that respiration cannot be performed in them as it is in man and animals similar to him; the absence of the osseous parts and diaphragm is sufficient to demonstrate this; and though, on the records of physiology, there are instances of the continuation of respiration after the mobility of the osseous parts had ceased, yet, as these were only instances of suffering nature, where the accompanying assistant, the diaphragm, still continued in full energy, physiologists ought, likewise, in examining the structure of the animals I am now to treat of, the Tortoise-tribe, to have suspected that this function was not performed in them as it is in us. Yet these hints given by this anomalous structure have either been neglected or made an improper use of, and the manner of their respiring remains in the greatest obscurity to the present hour. Before I proceed to show the present state of our knowledge on this subject, by giving the opinions of the celebrated anatomists and physiologists who have written upon it, I will just observe that, as the impossibility of respiration being performed in the frog-tribe, in the usual manner, consists in the absence of the ribs and diaphragm, so here the immobility of the whole bones of the trunk, and absence of the diaphragm, form the insuperable hindrance, and not a deficiency of solid parts as in the preceding; for a modification of the ribs and sternum here envelops the whole animal. The diaphragm, though said by some to exist, is really wanting. Blasius, however, asserts its existence, and describes it thus: "Diaphragma insigne admodum, oblique a pectoris anteriore inferioresque parte sursum adscendet, lateribus primo, hinc dorso firmiter adhærens; altiorem adeoque situm in posticis obtinet, quam in anticis, contra ac in homine, canibus bobus aliisque animalibus observamus, ubi anteriora sublimem majis locum habent posterioribus. Membranosum hoc totum notatur, similiter ac in avibus variis deprehendimus, nullis fibris carneis manifeste gaudens. Distinguit equidem thoracem a ventre inferiore, ast non sit in animalibus aliis: Pulmones enim cum hic sese in hoc magis, quam illo ventre exhibeant magne parte, diaphragmate haud includuntur, imo vix aliqua parte. Extendit se supra hepar partesque alias ipsi adsitas, usque ad vesicam urinariam cui valide adeo unitur tota superficie superiore ut non nisi magno artificio separari queat. Superius pericardio jungitur." But I am convinced he has taken the peritonæum for it. I have sought for it in vain, as well as other zootomists; neither Gotwald nor Wallbaum has observed it, and the French academicians, who dissected one near five feet long, say, that "la tortue a non seulement son écaille, qui lui sont lieu de thorax, absolument immobile, mais nous ne lui avons trouvé n'y de diaphragme, n'y d'autres parties qui puissent supplier à ce mouvement." This deficiency of the requisite mechanism for respiration has led some physiologists to explain this important function upon principles inconsistent with sound physiology, analogy, or experience. Perault attributes the expansion of the lungs, and consequent inspiration, to the elasticity of the membranes forming their cells; and the expiration to the compression of muscles, of which, he says, these animals have plenty. "Apparement," he says, "il est nécessaire de supposer que l'inspiration se fait par le ressort des ligamens durs et fermes qui composent les mailles qui ont été décrites: en sorte que lorsque les muscles qui peuvent comprimer le poumon viennent à se relacher, les ligamens s'étendent et élargissant les ouvertes de toutes les vessies augment la capacité de tout le poumon." Varnier boldly asserted that the whole process of respiration, both expiration as well as inspiration, was effected by the lungs themselves alone, by the means of their muscular texture, as a muscular network surrounded them, by which means they could respire by the alternate dilatation and contraction of the vesicules without the aid of the other instruments of respiration. He says, "Je parvins à me démontrer à moimême que le poumon de la tortue etoit entouré d'un réseau musculaire que par ce moyen ils étoit parfaitement irritable, qu'ils avoit une action propre, indépendente des autres agens de la respiration et qu'ils pouvoit inspirer par lui même;" and soon after adds, "le muscle du poumon de la tortue qui produit un mouvement convulsive," and then says that, "dans le tortue le poumon est cellulaire; les cellules se correspondent comme dans la grenouille; le muscle enveloppe toute la masse, et en se contractant la remue toute entière;" and concludes his memoire by saying, "le poumon est un organ actif; qu'il est le premier et le principal agent de la respiration, et que cette fonction dépend, comme dans les amphibies, de la dilatation et contraction alternative des vésicules qui determinent alternativement la contraction des muscles inspirateurs et expirateurs, et cela indépendamment de la volonté." Admitting the lungs to possess this muscular texture, which could not be perceived by Haller and the best anatomists, they would still be ill adapted to inflate by their own power. We learn, through the Transactions of the Royal Academy of Paris, that it was the opinion of Monsieur Tauvry that they breathed only in walking. "La tortue est enfirmée entre deux écailles immobiles, et elles n'a d'ailleurs aucun diaphragme qui puisse servir à une compression alternative des poumons. Dans cette difficulté d'expliquer sa respiration, Monsieur Tauvry s'est avisé d'en rapporter la cause au mouvement du marcher; quand la tortue est en repos, sa tête et ses pies sont retirés sous l'écaille supérieure, et la peau qui l'enveloppe entièrement est plissé, mais quand l'animal marche, il pousse au dehors sa tête et ses pies; sa peau s'étend, puisqu'elle est tirée par ces parties, et par conséquent elle forme intérieurement un plus grand espace, et c'est dans cet espace vuide que l'air extérieur est obligé d'entrer." This explanation, which is very anomalous with everything we know of this function in other animals, I put to the test of the following experiments, which proved it erroneous. I took the Testudo orbicularis in its contracted state, and wrapt it up in paper, binding it all round with bandages so fast, that the testa and sternum were brought so near before as not to admit the exit of the head. I then made an aperture in the paper opposite to its nose, and thus deprived of every motion, I placed it before the flame of a candle, yet I found not only that it blew the flame, but sometimes so strongly as nearly to extinguish it.

This experiment, though conclusive against the opinion of Tauvry, I strengthen by another; in this I kept the legs out, binding them very firmly under the sternum, the head being contracted as before, yet I still observed that it breathed, and as in the former experiment, sometimes with great force. The respiration, therefore, of the tortoise has no more connection with its other motions than that of other animals. But Morgagni, who was, as I have mentioned in the second dissertation, acquainted with the manner of respiring in frogs, which I have given in detail, supposes that the tortoise respires in the same manner; for, speaking of the frog, he says: "Inspiratio autem iis instrumentis per quæ inferior buccæ pars amplificata animal contracta ærem in pulmones compellit;" and then adds, "quin imo id ipsum, dum fluvialem quandam testudinem vivam inciderem, observavi invenique, totam eam partem quæ intra cavitatem mandibulæ inferioris est, multum posse extrorsum curvari ut hinc ær immitti posset, pulmones vero fibrarum rite firmari, ut hinc ær vicissim posset remitti." Notwithstanding the high reputation of Morgagni, I must dissent from the opinion of the tortoise respiring like the frog. I will not say that none of the genus do respire in this manner, as I have had no opportunity of examining any of the turtles. I wish to be understood as speaking of the Testudo orbicularis, my observations having chiefly been confined to this species, though I think I may say the same of the græca and palustris. Yet the opinion of this celebrated man is supported by Coiter and Varnier saying that, after the sternum is taken off, and the lungs are laid bare, the animal can still inflate them. But if, after the sternum was taken off, the peritoneum cut through, and the lungs laid bare, these appeared to Coiter and Varnier to inflate, this might not have proceeded from any power residing in the lungs themselves, nor from any air being impelled into them by the muscles of the throat, but by the parts in contact with them, as the neck before, and the muscles behind (which I shall soon describe), shortening them, in which case they would appear more distended, though the quantity of air within was not increased. The tortoises which I opened I never observed to inflate their lungs as the frogs do; nor did the anatomists mentioned by Valentini observe it, for they say, "Pulmones enim depressi remanebent, nec inflabuntur ab illa aeris attractione quod fieri potuisset tamen ab animali adhuc vivente licet capite truncato, quod ego subito, antequam aperiri, curarem, abscindi jusseram." Yet adds, "Vitalis autem haec testudo actiones habuit horæ spatio absque corde sed et absque capite; nam pedas movit ad tactum nostrum et sine eodem quoque retraxit." These are the opinions of the older anatomists; and amongst the moderns I know of none who have said anything on this subject. Being dissatisfied with them, I entered into the investigation by actual observation, and opened one for this purpose. The sternum being taken off with great care, the periosteum presents itself as a strong white membrane like parchment; when I had cut through this, I found many muscles inserted into it, particularly over the scapulæ and os pubis, which, in the contracted state of the animal, are not far asunder. Just above the os pubis it is connected to the peritonæum; by this means these bones, with their muscles, are enclosed as in a bag, having the peritonæum beneath, and the periosteum above; the scapulæ, and their connected bones and muscles, are enclosed in the same manner. The peritonæum being cut through, and the intestinal canal, liver, &c., removed, the lungs, consisting of two lobes, are seen covering nearly the whole of the testæ; they are cellular, as in the frog, and consist of two lobes, one on each side of the spina dorsi, each of which is subdivided into four or five indistinct lobules. The cellular texture of these is not uniform; the cells of the middle lobules being the smallest, and those of the last lobule the greatest; this lobule is likewise loose, not being tied down on the sides nor beneath, the rest are tied down to the spine. My attention was soon called to observe the structure and office of some muscles in the region of the flanks, which I observed often to be in motion, contracting and extending alternately, and though placed by the side of the hind legs, these were not moved by them. Further, they were placed at the end of the last lobule of the lungs, and they appeared to retain their irritability the longest. This was sufficient to lead me to conjecture that these might be the parts by which respiration in these animals was performed; and to see them act in their natural position I sawed off, in another tortoise, that part of the shell which covers them, and I then saw them constantly working. One was now placed nearly in a perpendicular direction, and another, or part of the same, was placed nearer the sternum, lying almost in a horizontal direction. The first in its contraction receded from the testa inwards, whilst the latter, in its contraction, observed a contrary direction. When I attributed to them the office of expirator and inspirator muscles, which I supposed them to perform, I was embarrassed, because I could not conceive how a muscle could be a constrictor with its convex side; yet when the expirator, by contracting, had receded from the shell inwards, it appeared, when viewed from without, to be concave. But this difficulty ceased as soon as I had opened the animal and dissected the parts, for I then found the following admirable contrivance of nature. This part is composed of two distinct muscles, with their risings and insertions quite different, yet firmly connected in the middle by cellular membrane. The first rises from the testa near the spina dorsi, and is inserted into the peritonæum; this is the constrictor of the lungs, or the muscle of expiration. The other is nearly spread over the whole cavity between the upper and under shell, where the hind legs are drawn in during the contracted state of the animal, being inserted into the margin of the testa above, and the margin of the sternum below. The places of insertion of these muscles, and their connection in the middle being known, there is then no difficulty in explaining why the muscle, while acting as a constrictor, appeared concave, as it was only the inspirator brought into that position by its antagonist; nor any difficulty in conceiving how they carry on the function of respiration; for the expirator being connected, as I have already said, to the testa below and to the peritonæum above, envelops in a manner the last movable lobule of the lungs; when, therefore, it contracts, it compresses this part of the lungs, and by that means expels the air; then ceasing to act, the other contracts, and draws the former with it, thus a vacuum is formed, into which the air rushes, as in the respiration of those animals which have a thorax.

To prove that this explanation was well-founded, and that the motions of these muscles were really those of respiration, I made the following experiment. I fastened on the nose of a tortoise a little valve made of white paper, which covered the nostrils, and with the assistance of a friend, I watched the motions of the soft parts lying within the hollow where the hind legs came out, and I found that these motions perfectly corresponded with the motions of the valve, which was put into motion by the expirations and inspirations of the animal. In this manner I conceive respiration to be carried on in the tortoise, without, however, meaning to extend this explanation to the whole of the genus Testudo, some families of which I have never yet had an opportunity to examine. These animals will therefore materially differ from those of the two preceding families in the mode of respiring; the air in them being driven into the lungs by the muscles of the throat, which act like a pair of bellows, whilst in these it is performed by the lungs following the motions of their containing parts, and they will therefore differ from the animals having a thorax chiefly in the form and situation of the parts. Respiration is not, I think, the only office of the muscles which I have just described; they are closely connected to the bladder, and to them, I think, this animal is indebted for the power it possesses of sucking in water by the anus, as I mentioned in my last dissertation; but this investigation I must leave to another time.

It will thus be seen, while this close observer fully realized that respiration in the turtle was not effected by deglutition, but by muscles of expiration and inspiration situated in the flank spaces, yet, failing to recognize the true office of the anterior muscles, his conception of the respiratory process was necessarily imperfect and insufficient, and to this, no doubt, must be ascribed the neglect into which his views have fallen.

In 1819 appeared the most important contribution to the literature of the subject, the monograph of LUDOVICUS HENRICUS BOJANUS, on the Anatomy of the Testudo Europaeae. This work being purely anatomical, we have no means of judging as to the author's knowledge of the muscular apparatus concerned in respiration, except by the nomenclature he adopts, and some details of description. The inspiratory muscle and the posterior belly of the expiratory muscle are grouped as abdominal muscles, and described as the obliquus and transversus-abdominis, while the anterior belly of the expiratory muscles, under the name of diaphragmaticus, is thus referred to: "A corpore vertebrae dorsi quartae et tertiae et a costa tertia oriundus, triplici fasciculo complanato, divergentibus eundo; quorum bini ad marginem internum pulmonis, sui lateris, descendunt eique agglutinantur; tertius vero supra pulmonis anterius extremum revolutes ad peritoneum (a pulmonum facie inferiore versus cardiam et hepar deflectens) desinit." It is, no doubt, probable that these names have been determined by supposed homologies of the muscles, and yet we may reasonably conclude that Bojanus had not perceived any relationship between the diaphragmaticus and the transversus abdominis, and did not realize that the broad fibrous membrane extending between them was their common tendon. This conception is essential to the full realization of the respiratory process.

G. DE CUVIER, bearing in mind the type of batrachian respiration, regards the alternate contraction and dilatation of the throat as movements of deglutition of air, and thinks them a sufficient and the only means by which inspiration is effected. The expulsion of the air from the lungs he refers to the agency of two muscles in the flank, the obliquus and transversus of Bojanus, at the same time calling attention to the fact that Townson has erroneously attributed to one of these (the obliquus) the function of an inspirator. He thinks also that the analogues of the quadratus lumborum and the rectus abdominis, by compressing the viscera, may assist in expiration. In his Leçons d'Anatomie Comparée, vol. vii. p. 216, Duvernoy's edition, 1840, we find his opinion in detail.

"Le même mécanisme est mis en jeu dans les chéloniens. La déglutition de l'air est le seul moyen dont ils puissent se servir pour faire entrer ce fluide dans leurs poumons. Ils dilatent et contractent leur gorge alternativement, ayant la bouche fermée, absolument comme les batraciens et par les mêmes puissances. Il est expulsée par deux pairs de muscles analogues à ceux du bas-ventre des animaux précédents. Ces muscles remplissent l'intervalle postérieur du sternum et de la carapace, dans lequel se replient les extrémités pelviennes dans l'état de repos, et c'est à cet endroit qu'on aperçoit dans les chéloniens les mouvements de contraction et de dilatation qui, dans les mammif. res, se voient dans toute l'étendue du ventre.(1) La première paire ou l'externe résemblent à l'oblique descendant, elle s'attache à tout le bord antérieur du bassin, à la carapace et au sternum, et s'étend dans tout l'intervalle postérieur de ces deux parties. L'interne ou l'analogue du transverse est composé de fibres transversales s'attachent supérieurement à la moitié postérieur de la carapace près des vertèbres, descendent en dehors des viscères, les enveloppent et viennent aboutir inférieurement à une aponevreuse moyenne. Celle-ci passe en partie sous la face inférieure de la vessie, et doit servir à la vider lorsque ces muscles se contractent. Ils ne comprennent immédiatement qu'une petite portion des poumons ; mais leur action s'exerçant plus fortement sur les viscères du bas-ventre, ceux-ci pressent à leur tour les premiers organes et en expulsent l'air. Les muscles analogues du quarré des lombes et du droit abdominal qui ont été décrits (t. i, pp. 488, 489) doivent aussi comprimer les viscères abdominaux, et par leur moyens les poumons. Les chéloniens qui ont leur cavité viscérale divisé par le pleuro-péritoine à la manière de celle des oiseaux, ont une de ces cloisons celle qui descend de la partie antérieure du bouclier dorsal, au devant du foie, constituée comme un diaphragme par des fibres musculaires et aponevretiques. Bojanus décrit un muscle diaphragmatique pair composé de fibres musculaires épanouies de chaque côté sur cette cloison, que nous avons fait connaitre comme une sorte de diaphragme (t. iv, 2d partie, p. 651). Son action, quoique faible, peut contribuer à l'expiration en comprenant les poumons.

"Peut être que les poumons se contractent aussi par une force propre que réside dans le réseau tendineux qui entre dans leur composition (Art. 11, de cette Leçon, p. 130).

"N. 1. Je crois l'avoir fait connaître le premier (Bull. de la Soc. Phil. an. xiii, No. 97, p. 279) en démontrant, contrairement à l'opinion de Townson, que les muscles du bas-ventre sont l'un et l'autre des muscles expirateur. Et cependant c'est à cet auteur qu'on attribue l'explication que j'ai donnée en montrant l'inexactitude de la sienne."

Dumeril et Bibron, vol. i. p. 176, 1834, described briefly the mechanism of respiration in chelonians thus: air enters the buccal cavity through the nose, then the fleshy tongue is applied to the posterior nares so as to close them, and the mylo-hyoid floor of the mouth contracts, to force the imprisoned air into the lung. A succession of such acts fills it.

Before entering upon the details of description, it may be well to premise, that this anatomical section of our paper is intended mainly as an exposition of the muscular and neural apparatus by means of which the movements of air to and from the lungs are effected in chelonians, and while, to render the subject more intelligible, we shall rehearse the general anatomy of the organs of respiration, we shall avoid all questions of structure or function irrelative to the point of inquiry, referring the reader desirous of such knowledge to the more general works on comparative anatomy.

Underlying the floor of the mouth, and embracing with its cornua the sides of the pharynx posterior to the jaw, is the hyoid apparatus, or the tongue-bone, Fig. 1,

Fig. 1.

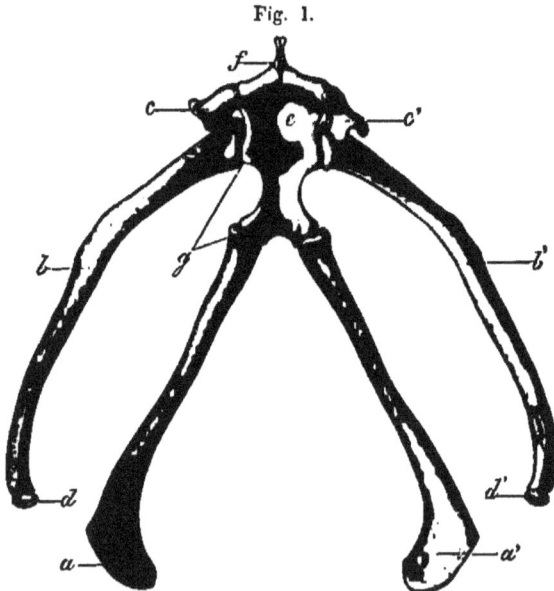

Fig. 1. a, a', lesser cornua; b, b', greater cornua; c, c', cartilaginous processes; d, d', ossicles for attachment of suspensory ligaments; e, body of bone; f, fenestrum, closed by cartilage; g, articulations of cornua with body.

an instrument conspicuous for the part it has evidently played in fixing upon its possessor the batrachian type of respiration. It consists of an elongated body, excavated for the lodgment of the larynx and upper rings of the trachea, and of a cartilaginous process and two bony cornua on each side, connected with it by movable articulations. To the extremity of the anterior or major cornu is attached a knob or ossicle, for the reception of the suspensory ligament. This ligament arises from the mastoid process of the temporal bone of the cranium, and forms the fulcrum on which the apparatus swings backwards and forwards, moved by alternate contractions of the genio-hyoid and omo-hyoid, and other muscles of the neck. The hyoid bone, in its movements, carries with it the glottis, and removes it from obstructions during inspiration. The larynx, Fig. 2, A and B, consists of a largely-developed cricoid cartilage and two arytenoid cartilages. The cricoid rests in the bowl of the hyoid bone, is somewhat helmet-shaped, and has on its under surface a visor-like oval fenestrum. This fenestrum is covered by membrane, and is traversed from side to side by the chiasm of the superior laryngeal nerves, of which we shall speak more fully hereafter. Superiorly the cricoid presents an oval opening, filled in by membrane, upon which rest the arytenoid cartilages, one on either side, with the glottic slit

A
Fig. 2.
B

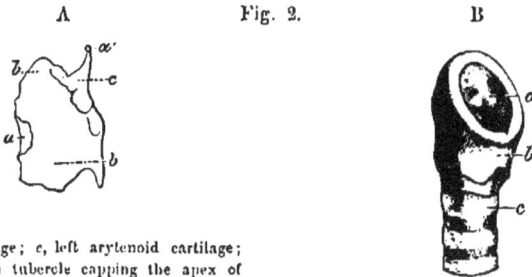

b b'. Cricoid cartilage; c, left arytenoid cartilage; a', cartilaginous tubercle capping the apex of the arytenoid cartilage; a, oval fenestrum of cricoid, filled in with membrane.

b. Cricoid cartilage; a, superior opening; c, trachea.

between them. The arytenoid cartilages, Fig. 2, A, c, are two irregularly triangular solids, opposing flat surfaces to each other, their bases incorporated with the superior cricoid membrane, and their apices extending vertically, and terminating in a small cartilaginous tubercle. They are the framework upon which the crico-hyoid and crico-arytenoid muscles act, in closing and opening the glottis. The trachea, smaller in diameter than the cricoid bulb, descends the neck, inclining to the left side, until opposite the margin of the shell it divides into two bronchi, which, curving right and left, open free into the corresponding lungs, at the under and inner edge, a little behind the anterior extremity. The lungs are two wedge-shaped sacs, the base of the wedge being anterior. They lie in contact with the vault of the dorsal shield, and are separated from each other by the large retractor muscles of the neck, the bloodvessels, and nerves. They are anterior and above the peritoneal sac, except the posterior pointed extremity, which projects into that cavity, carrying with it a covering of peritoneum. The walls of the lungs being elastic lend aid to the act of expiration, but as they give no evidence of muscular fibre to mechanical or galvanic stimuli or to the microscope, it is impossible, for this and for other reasons, to suppose with Varnier that they possess any intrinsic power to assist in the act of inspiration.

The turtle which has served us for most of our experiments, is Chelydra Serpentina, the well-known Snapping Turtle of the United States. Selected at first from the facility with which we could procure fine specimens, we soon found that its well developed muscular system and its exposed flanks admirably fitted it for the study of respiratory myology, while its middle rank among Testudinata led us to expect, in its organization, more striking ordinal characters than we would find in the extreme marine or terrestrial families. We have therefore adopted Chelydra

2

Serpentina as our typical turtle, and will describe in detail the apparatus of respiration as we find it in this species, noting, subsequently, the modifications of structure existing in the different genera that we have had the opportunity to examine. In all turtles we have found the general plan of the respiratory apparatus constant, an inspiratory muscle in each flank, and an expiratory muscle with four bellies, two anterior and two posterior, connected by a broad membranous tendon, inclosing the viscera and capable of compressing them against the under surface of the dorsal shield. The discrepancies characterizing different genera principally affect the origin of the anterior belly of the expiratory muscle; these may naturally be arranged in two groups, those in which the origin is anterior (about the second rib), (see Fig. 5) and extends nearly across the width of the shield, and those in which the origin is posterior (about the third or fourth rib), and in extent more limited. The specimens we have had the opportunity to examine are too few to enable us to determine whether this structural diversity can be received as an element in determining generic rank. We will content ourselves, therefore, at present, with the description of each specimen, including a brief notice of its habits and shell-measurements, which may serve as a nucleus for future and more extended observations.

Chelydra Serpentina is a carnivorous turtle living in the water, under bank-caves, or at the bottom of streams, and yet capable of moving over the land with facility. The under surface of the body is much exposed, the plastron being small and cruciform, and connected with the carapace by a narrow bridge, which widens to join the fourth, fifth, and sixth ribs. The flank spaces are large, flat, and unprotected, and the extremities incapable of complete retraction under the shell. The height of the trunk compared with the width and length of the carapace is as one to three and three and a half.

Carefully watching the animal while breathing, we notice synchronous movements of the trunk, of the throat, and of the glottis within the mouth. With the first element of the respiratory act, expiration, the glottis opens, the hyoid apparatus descends and widens, the shoulders sink and the flanks become increasingly concave; then follows immediately the inspiratory effort, the glottis remaining open, the throat narrows, the flanks become tense, and the shoulders are pushed forwards as the act culminates; afterwards the muscles relax, the glottis closes, and the creature is at rest until again impelled to renew the air in its lungs, when the same sequence of expiration, inspiration and pause is repeated.

We shall follow the order of the elements of the respiratory act in describing the apparatus by which it is effected. And first, of the muscles of expiration. For the purpose of dissection, it is desirable to place the animal upon its back and fix it, by extending and securing its head, tail, and extremities. Separate with a saw the bony bridges connecting the plastron with the carapace, and sweeping a knife close to the inner surface of the former, divide from before, backwards, the deltoid, pectoral, pelvic and flank muscles, the acromial articulations posterior to the first pair of sternal bones, and the loose cellular bands binding the visceral sac to the middle line. This permits the removal of the plastron. Drawing the shoulders forward, cut the ligaments, holding the scapulæ to the spine anterior to the first rib, which loosens the entire muscular and bony mass, and facilitates its detachment. The section should be made with the lung partially inflated, to secure from injury the anterior belly of the expiratory muscle, which lies in contact with the posterior surface of the serratus magnus. The further removal of the tissues of the flank, and their careful separation from the posterior belly of the expiratory muscle to which they are adherent, completes the exposure.

Looking at the result of our dissection, we find a tendinous and muscular sac occupying the dorsal shield, filling its entire width in the middle and most of its length; its general form is cordate, the apex dipping into the pelvis, and its anterior notch giving place to the heart and the muscles and vessels of the neck. Much the larger portion of the sac visible is tendon (Fig. 3, g, g'), and has hitherto been regarded as peritoneum, but a closer scrutiny would have revealed its fibres gathering from all sides towards an oval centre, in which they are inseparably interwoven. The tendon in many places can be lifted from the peritoneum, by which it is lined. Curving around its anterior and posterior borders are muscular fringes (Fig. 3, d, d' and e), the fibres running from the carapace in lines parallel to the long axis of the trunk. These are the anterior and posterior bellies of the expiratory muscle, the diaphragmaticus and transversus abdominis of Bojanus. These muscles are inserted into the common tendon, and in contracting compress the contained viscera against the shell and expel the air from the lungs. Dividing the tendon through its middle from side to side, and removing the abdominal organs and permitting the lungs to collapse, we are enabled to obtain a satisfactory view of the origin of these muscular bellies from within.

The posterior belly arises from the pelvic fascia from a point opposite the anterior third of the ilium backwards to the spine, from the eighth vertebra, and by tendinous fibres from the carapace as far as the sixth rib, the line of origin slowly leaving the spinal column as it reaches forwards. Turning outwards at an obtuse angle, after joining the sixth rib, the muscle follows its posterior edge until near its extremity, where it inclines forwards and terminates at the fifth rib as it joins the marginal plates, a point corresponding very nearly with the pelvic end of the suture connecting the carapace and plastron.

From this sigma-shaped origin the fibres curve backwards and downwards, embracing the abdominal viscera, and unite with the tendon below, forming a regular and well-defined line, varying in position as the muscle is contracting or at rest. Fig. 3, c, represents the lungs distended and the muscle relaxed. This belly, considered by itself, is a strong membranous muscle, somewhat triangular in shape, the apex being at the edge of the shell, and the base at the pelvis. In a turtle weighing sixteen pounds, the fibres at the apex measured one-half inch in length, while in the middle and at the base they measured respectively five and one-half and four inches.

The anterior belly arises from the vertebral margins of the second and third intercostal spaces, from the second costal arch, from the second rib along two-thirds of its length, and across the carapace in a line curving backwards and outwards, from the third and fourth ribs, near their junction with the marginal plates. It will thus be seen that the outermost origins of the anterior and posterior bellies closely approximate above the plastron where it meets the upper shield, while at the middle line of the body the origins are separated by the distance of the eighth from the third vertebra. The fibres composing the anterior belly are close and firm for the outer

Fig. 3.

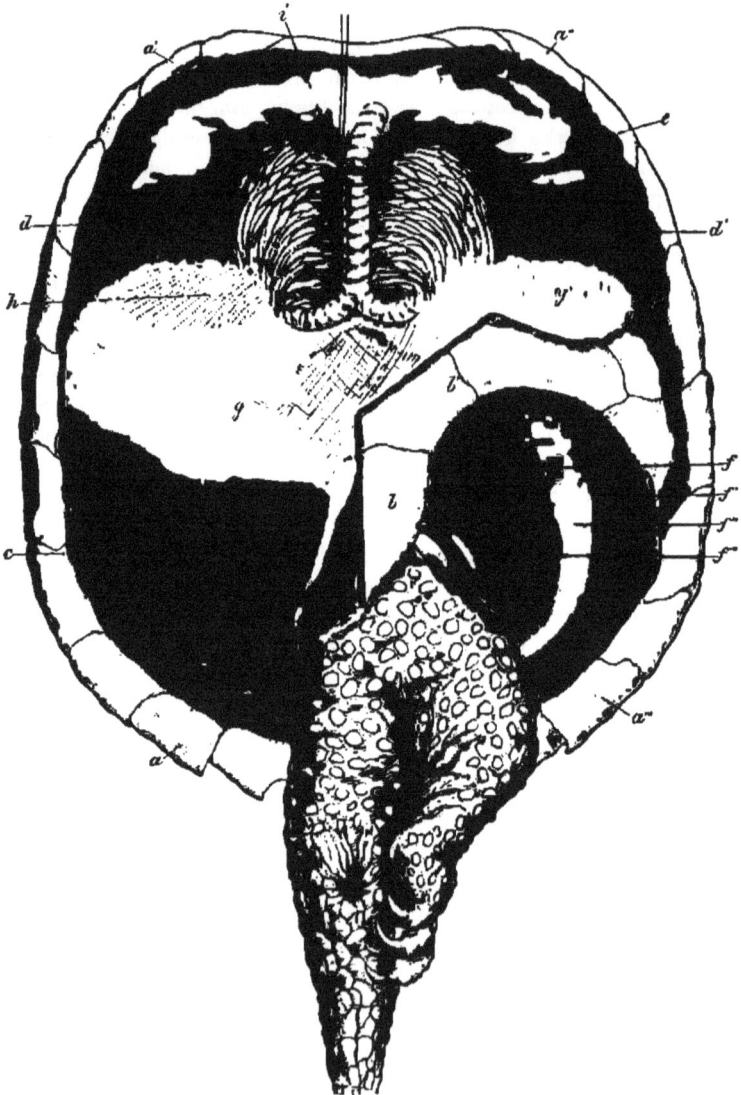

Fig. 3. Muscles of inspiration and expiration.—*a a′ a″ a‴*, margin of carapace; *b b′*, portion of plastron in position; *c*, posterior belly of the expiratory muscle on the right side; *d d′*, anterior bellies of the expiratory muscle; *e*, reticulated portion showing the lung beneath; *f f′*, inspiratory muscle of the left side; *f″*, central tendon; *f‴*, tendinous ligament; *g g′*, tendon of expiratory muscle; *h*, muscular fibres beneath the tendon *y y′*, and attached to the lung.

half, while those capping the inner portion are fewer in number and reticulated, permitting the lung to be seen through their interspaces. (See Fig. 3, e.) The part of the muscle which arises from the vertebræ covers that triangular surface of the lung which looks towards the interpulmonary notch, while that of costal origin spreads over the anterior face of the lung, sheathing its entire thickness when the organ is fully inflated. A few of the fibres capping the anterior and superior extremity of the lung continue their course over the under surface of that organ, spreading fan-like towards its outer edge, and being inserted into its adherent peritoneal covering. They are represented by the dotted lines (Fig. 3, h). These fibres are much more largely developed in some other genera, and seem to have the power of drawing the lung in towards the spine, and keeping it well under the viscera when compressed during expiration.

The inspiratory muscles (Fig. 3, f') are to be sought for in the flank spaces at the under and posterior portion of the trunk, into which the hind limbs of the animal are drawn during repose. There is one muscle in each flank, superficial, and readily displayed, by reason of the loose cellular connection it has with the tissues concealing it. Turning aside the skin and fascia loaded with adipose matter, as it often is in this locality, we at once expose this beautiful sheet of muscular fibre, during contraction, stretching like a drum-head over the entire space, and fitting closely its irregular boundary. Through its centre, from before backwards, runs a flat tendon (Fig. 3, f''), averaging in width one-sixth of the breadth of the muscle, and receiving throughout its length, on both sides, the insertion of fibres. It is usually a single band, but in several specimens we found it irregularly double, being divided by islets or patches of muscular fibre. In some form, however, it exists in all Chelydra, and constitutes a striking feature of the muscle, its white pearly hue contrasting boldly with the crimson fringes between which it is placed. In some families it loses its significance, dwindling to a central raphé, or more rarely is absent altogether. The direction of the muscular fibres is transverse, especially in the anterior part of the space; behind and outside of the tendon they diverge to accommodate themselves to the circular sweep of the carapace. Being attached to no other mobile part than the central tendon, we may consider that as their insertion; their origin embracing the entire circumference of the space. Beginning with the posterior sternal bone, we may trace its fibres coming from the inner edge of the plastron, where it curves around the flank, from within the marginal plates of the carapace, from the fascia filling the space posterior to the sacrum, and along the pelvic muscles from a ligament, the counterpart of Poupart's ligament in man, stretching between the ilium and pubis. The fibres arising from the anterior end of this ligament underlie the lowest fibres from the plastron, and give to the latter a falciform appearance, represented in (Fig. 3, f'''). On the upper side, the inspiratory muscle is attached by cellular tissue to the posterior belly of the muscle of expiration, and by the contraction of this latter muscle during the expulsion of air from the lung, is carried downwards and forwards into a strongly concave position, most favorable for its own subsequent effort.

The capability of the turtle to hold the air in its lung at will, or when subjected to great external pressure, as must constantly occur in marine species,

Fig. 4.

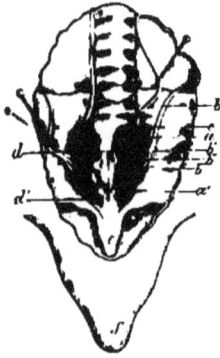

Fig. 4. Glottic muscles and nerves. —a a', crico-hyoid; the muscle overlying it is the crico-arytenoid; b, superior laryngeal nerve; b', communicating branch; b'', branch to crico-arytenoid; b''', branch to crico-hyoid; c, recurrent laryngeal; d d', glottic slit; e, point of hyoid bone; f, tongue.

is determined by two pairs of muscles situated about the glottis, and controlling its movements. These are the crico-arytenoid and the crico-hyoid of Bojanus; to the former is intrusted the opening of the glottic lips, while the latter, acting as a sphincter, serves to close them. The crico-arytenoid (Fig. 4) lies beneath the mucous membrane, superficial to the crico-hyoid, and crossing it nearly at right angles. It arises from the sides of the cricoid cartilage anteriorly, and is inserted into the body and vertical limb of the arytenoid as far as its extreme point.

The crico-hyoid (Fig. 4, a a') arises from the body of the hyoid bone anterior to the depression for the larynx, its middle resting upon and exterior to the arytenoid cartilage. It is inserted into the cricoid cartilage at its anterior raphé. The muscles of the two sides approximate each other at their origins, and interlace at their insertions, forming an elliptical muscle surrounding the rima glottidis.

Our opportunities for studying the arrangement of the respiratory muscles in other turtles than Chelydra have been limited to the representatives of two families, Chelonioidæ and Emydoidæ.

Among the Chelonians we have examined but one species, Chelonia mydas, the Green Turtle of the Atlantic Ocean. Its habits are entirely aquatic, seeking the land only for the purpose of depositing its eggs. The body is flat, the under surface well covered by the plastron, leaving, however, naked flank spaces, as in the snapper. The union between the plastron and carapace extends from the second to the seventh rib. The inspiratory muscles occupy the flanks, arising a half inch or more within that part of their boundary which is formed by the plastron. The central tendon exists, and is wide and irregular, and extends the whole length of the muscle.

The origin of the expiratory muscles is similar to that found in Chelydra; the muscular bellies are shorter, however, and the common tendon broader and longer in accordance with the shape of the turtle.

The dimensions of the shell are—length, 38 inches; width, 28 inches; elevation, 13 inches.

Among the Emydoidæ we have examined individuals from eight genera, and find them to present considerable variations in the origin of the anterior belly of the muscle of expiration. And as these differences seem to characterize groups in harmony with the subdivisions of Agassiz, founded on minor differences of form observed in this family, we shall follow his classification in their description.

The first subdivision suggested by this distinguished observer, and styled Nectemydoidæ, is thus characterized: "The body is rather flat. The bridge connecting the plastron and carapace is wide, but flat. The hind legs are stouter than the fore legs, and provided with a broad web, extending beyond the articulation of the nail joint. The representatives of this group are the largest and most aquatic of the whole family." Of the genera included in this sub-family we have observed four: Ptychemys, Graptemys, Malacoclemmys and Chrysemys.

Fig. 5.

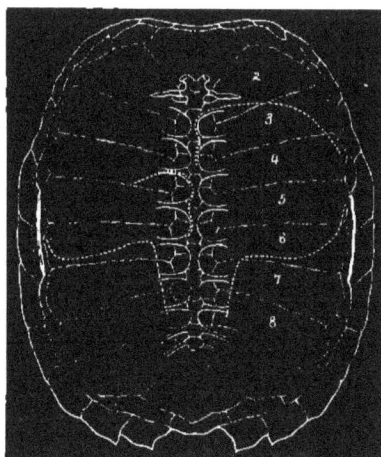

Fig. 5. Diagram of carapace of turtle, showing with the dotted lines the two principal types of origin of the expiratory muscle. The left side of the diagram shows the line of origin in the most aquatic species. The right side that of the most terrestrial. The numbers indicate the ribs.

Ptychemys rugosa, Ag.—The inspiratory muscles are found in the flanks as usual, but they have no central tendon, a simple line or raphé marking the junction of the converging fibres.

The anterior belly of the expiratory muscles arises from the vertebral margin of the fourth and fifth intercostal spaces and from the surface of the fourth rib near its posterior edge for a distance one-third its length. From this right-angular origin the fibres diverge, expanding over the upper and anterior surface of the lung, to join the common tendon at the anterior and inferior pulmonary margin. The fibres extending back on the under surface of the lung, as indicated by the dotted lines (Fig. 3, *h*), are numerous and large in this species, and seem almost to foreshadow the muscular separation between the thoracic and abdominal viscera in higher vertebrates.

The posterior belly in its origin presents no variation from that of the Snapper. Its outermost fibres, however, are much developed, forming a muscular band which reaches forwards nearly as far as the anterior junction of the carapace and plastron. The dimensions of this turtle are—length, 11 inches; width, $8\frac{1}{10}$ inches; and elevation, 5 inches.

Ptychemys mobiliensis.—Shell measurements. Length, $13\frac{1}{2}$ inches; width, $9\frac{1}{10}$ inches; elevation, $6\frac{1}{10}$ inches. The general shape and appearance of this turtle resembles P. rugosa. The anterior and posterior extremities of the bridge connecting the plastron and carapace are much more strongly involute than we have observed in any other species. When the shell is separated, they appear like four projecting keels directed towards each other, the front ones looking inwards and backwards, and the posterior ones looking inwards and slightly forwards. The concavities thus formed before and behind, external to the keels, lodge projecting portions of the lungs. The anterior and posterior keels projecting into the space usually occupied by the air sacs, deeply indent them, and cause them to present a lobulated appearance, which they retain when removed from the shell. Besides these four large indentations, there are smaller ones, in the edge of the lungs, one in front and two or more between the keels. To the inner side and behind the posterior keel lies the large posterior lobe occupying chiefly the flank space immediately above and in front of the inspiratory muscle. The reticulations of its interior structure are much larger and more coarsely marked than those of other parts of the lung.

The anterior belly of the expiratory muscle arises from the vertebral margins of the third and fourth intercostal spaces, and from the carapace in a line diverging at an angle of 30° from the spine for the space of two inches, crossing the fourth rib. From this origin the fibres cover the front of the lung, the anterior and interior ones being distributed as in P. rugosa, and owing to the intrusion of the anterior keel upon the lung, the external fibres are displaced, so to speak, with the portions of lung to which they belong, which portions lie immediately back of the ridge or keel so often referred to. The largest band of those lateral fibres finds its way between the two lobules of the lung which lie first and second in order of succession behind the ridge. The posterior belly arises from the pelvic fascia, from the eighth and seventh vertebrae, and from a curved line whose convexity looks forwards, and which terminates in front of the posterior projecting keel about two and a half inches above the posterior angle of junction of the carapace and plastron. This line is rendered more sharply convex at its external third by the projection inwards of the keel alluded to. The muscular fibres curve around the posterior part of the lung. The inner ones for half the width of the muscle are about two and a quarter inches long; and from this point they increase gradually in length to the external edge, where they are longest, extending forward in a tongue-like band about five and a half inches. In the single specimen examined we found on the left side a few additional fibres reaching forwards and inwards at least two inches beyond the main body of the muscle. The inspiratory muscle arises as in P. rugosa. It has a linear central tendon, bifid at its posterior extremity, the shorter terminating arm being external. Into the tendon and its branches the muscular fibres are inserted as in other species.

In *Graptemys geographica*, Ag., the inspiratory muscle is, in its general features, the same as described in other species, and differs only in not having even a central raphé, the convergent fibres interlacing at the middle of the muscle in an imperfect network which serves to replace the tendon usually found in this situation. The anterior belly of the expiratory muscle arises from the vertebral margin of the third and fourth intercostal spaces, and continuously from the costal margin of the third space nearly its entire circumference, and from the surface of the fourth and fifth ribs. The lines of origin diverge at an angle of 30° from the anterior margin of the third intercostal space, and in this specimen, the inner line bordering the spine measures three-fourths of an inch, and the outer one, stretching along the intercostal space and across the shield, one inch and three-eighths; from this origin the fibres spread over the anterior part of the lung and are inserted into the common tendon and into the peritoneal covering of its under surface. The posterior belly is similar to that of P. rugosa, the muscular band underlying the bridge which joins the carapace and plastron being somewhat wider. The dimensions are—length, $8\frac{1}{4}$ inches; width, 6 inches; elevation, $3\frac{1}{4}$ inches.

In *Malacoclemmys palustris*, Ag., the inspiratory muscle is the same as in Geographica. The anterior belly of the expiratory muscle arises from the third and fourth spaces at their vertebral margins, and from a line running across the shield to the fourth rib, diverging at an angle of about 70°.

The posterior belly is like that of geographica. The dimensions are—length, 7 inches; width, $4\frac{3}{4}$ inches; elevation, $3\frac{4}{8}$ inches.

Chrysemys picta, Gray.—Inspiratory muscles as in E. terrapin. The anterior belly of the expiratory muscle arises from the vertebral margins of the third and fourth intercostal spaces and a slip from the fifth, and from across the carapace to the junction of the fourth and fifth ribs, the divergence being about 30°.

The posterior belly as in geographica. Dimensions—Length, $4\frac{2}{8}$ inches; width, $3\frac{1}{4}$ inches; elevation, $1\frac{5}{8}$ inches.

Of the second and third subdivisions we have examined no specimens. The fourth, Clemmydoidæ, is characterized by "their more arched though elongated form, and the more compact structure of their feet, the front and hind pairs of which are more nearly equal, and their toes united by a smaller web; they are less aquatic and generally smaller than the preceding." Of these we have dissected representatives of three genera, Nanemys, Calemys and Glyptemys.

In *Nanemys guttata*, Ag., the inspiratory muscle presents no peculiarities. The anterior belly of the expiratory muscle arises from the vertebral margins of the second and third intercostal spaces and from part of the fourth, and from the posterior edge of the second rib as far as its extremity; from this extensive origin the fibres descend over the lungs, covering the front and anterior part of their lateral walls.

The posterior belly resembles that of the Snapper. Dimensions are—length, $4\frac{1}{16}$ inches; width, $2\frac{4}{8}$ inches; elevation, $1\frac{7}{8}$ inches.

In *Calemys Mühlenbergii*, Ag., the muscles are the same as in guttata. Dimensions—length, $3\frac{5}{8}$ inches; width, $2\frac{4}{8}$ inches; elevation, $1\frac{1}{4}$ inches.

In *Glyptemys insculpta*, Ag., the muscles are the same as in guttata. Dimensions—length, $4\frac{9}{16}$ inches; width, $3\frac{7}{16}$ inches; elevation, $1\frac{5}{8}$ inches.

In the fifth subdivision, Cistudinina, "the body is remarkably short and high, slightly oblong, and almost round. The plastron, which is movable upon itself and upon the carapace, as in the Evemydoidæ, is also connected with the carapace by a narrow bridge; but the feet are very different, the toes, as in the Testudinina, being nearly free of web. Their habits are completely terrestrial." Of this subfamily we have examined one species, Cistudo Virginea, Ag. The flank spaces in which the inspiratory muscles play are extremely deep, owing to the high carapace. The amount of muscular fibre is relatively greater than in the other turtles, and the central tendon is narrow, and irregularly triangular in shape. The

3

anterior belly of the expiratory muscle arises from the vertebral margins of the second and third intercostal spaces and from the second rib throughout its length.

The posterior belly is like in origin to that of other Emydoidæ; the muscular fibres are longer, however, and terminate squarely in the tendon, as does also the anterior belly.

For convenience of reference, we have thrown into a tabular form the measurements and muscular origins of the above genera.

Species.	Mode of Life.	Dimensions in inches.			Origin of Respiratory Muscles.
		L.	W.	E.	
Nectemydoidæ.					
Ptychemys rugosa . . .		11	$8\frac{1}{5}$	5	4th and 5th vertebræ.
" mobiliensis		$13\frac{3}{4}$	$9\frac{4}{16}$	$6\frac{1}{6}$	3d and 4th "
Graptemys geographica	Aquatic.	$8\frac{1}{2}$	6	$3\frac{1}{4}$	3d and 4th "
Malacoclemmys palustris		7	$4\frac{3}{4}$	$3\frac{1}{2}$	3d and 4th "
Chrysemys picta . . .		$4\frac{1}{4}$	$3\frac{1}{4}$	$1\frac{1}{4}$	3d and 4th "
Clemmydoidæ.					
Nanemys guttata . .		$4\frac{1}{16}$	$2\frac{3}{4}$	$1\frac{0}{16}$	2d and 3d "
Calemys mühlenbergii .	Less aquatic.	$3\frac{1}{4}$	$2\frac{1}{4}$	$1\frac{1}{4}$	2d and 3d "
Glyptemys insculpta .		$4\frac{0}{16}$	$3\frac{1}{16}$	$1\frac{3}{8}$	2d and 3d "
Cistudininæ.					
Cistudo virginea . . .	Terrestrial.	$4\frac{3}{8}$	$3\frac{3}{8}$	$2\frac{3}{8}$	2d and 3d "

A glance at the table will show that in the most aquatic species of Emydoidæ the origin of the anterior belly of the muscle of expiration is from nearly the middle of the shell; while in the less aquatic and terrestrial genera it is from the forward part, and much more extensive. This arrangement is too uniform to be passed by unnoticed, although our facts are so few that we cannot form any conclusions as to its generic meaning. Whether the same diversity of origin exists in the genera of other families, and bears a similar relation to their family rank, and also whether this origin is modified during the development of the turtle, we must leave for future inquiry.

The neural apparatus of respiration in Chelonians, as in the Mammalia, consists essentially of the nervus vagus supplying the larynx, of spinal nerves distributed to the respiratory muscles of the trunk, and of the medulla oblongata, the common centre through which the synchronous movements of the glottis and of the flanks are incited and controlled. Between the ganglionic enlargements supplying the upper and lower extremities, the spinal cord is attenuated, the nerves coming from this region being restricted, by the existence of a bony thorax, almost entirely to those concerned in the movements of respiration. The disposition of the trunks of these nerves closely resembles that of the intercostals in man. Escaping from the spinal canal at the intervertebral foramina they traverse the carapace in parallel lines between the ribs, giving off branches from time to time to their appropriate muscles. By dissection and by mechanical irritation of the peripheral end of the cut nerve, exciting contraction of different fibres, we have determined that the filaments finally distributed to the expiratory muscle are derived from the first, second and third dorsal nerves for the anterior belly, and from the fifth, sixth and seventh for the posterior belly. The sixth and seventh nerves are also the sources of supply to the muscles of inspiration, the seventh being distributed over the inner or pelvic side, and the sixth to those fibres connecting the central tendon and carapace. Section of the medulla spinalis in the cervical region effectually intercepts communication between these nerves and their usual source of excitation. Under these circumstances the muscles of the trunk remain at rest, although the movements of the glottis indicate that the creature feels the respiratory need. These glottic movements continue normal even after the further section of both pneumogastrics.

The respiratory nerve of the larynx, the par vagum, emanating from the medulla oblongata, passes out of the cranium at the posterior jugular foramen, and courses down the neck within the sheath of the cervical vessels. Soon after leaving the skull, it gives off the superior laryngeals, and low down in the neck, opposite the aorta, the inferior laryngeal, the two branches that interest us at present. The superior laryngeal (Fig. 4, b), soon after separating from the parent nerve, approaches the major cornu of the hyoid bone, and under shelter of its posterior border, follows it closely to its junction with the body, then winding spirally forwards, it crosses the articulation, and runs along the margin of the excavation in close proximity with the larynx. In this position it gives off three principal branches. 1st. A communicating branch (Fig. 4, b'); 2d. A branch to the crico-arytenoid, or opening muscle of the glottis (Fig. 4, b''); and 3d. A branch to the crico-hyoid or glottic sphincter (Fig. 4, b'''). The communicating branch (Fig. 6) is a relatively large nerve, but has hitherto escaped observation; it is easily brought into view by dividing the trachea and lifting it forwards. It passes beneath the larynx directly from side to side, traversing the membrane of the cricoid fenestrum, about its middle. It is composed of fibres derived in part from each of the superior laryngeal nerves, which cross each other, to be distributed to the glottic muscles of the side opposite to that from which they originate. This remarkable nerve, we believe, furnishes the only known instance in nerve anatomy of an extracranial chiasm.

Fig. 6.

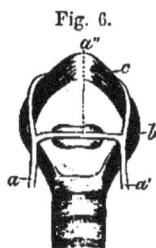

Fig. 6. The intercommunicating nerve seen from below.—a a', superior laryngeal nerves; a'', intercommunicating nerve crossing the fenestrum of the cricoid cartilage; b, crico-arytenoid muscle; c, crico-hyoid muscle.

Some few filaments penetrate the cricoid membrane, to be distributed to the mucous membrane of the larynx, and are doubtless sensitive fibres. At page 20 of the physiological section, will be found the experiments by which we have determined the function of this intercommunicating nerve.

The second and third branches present no peculiarities; they penetrate the muscles and are lost to view. Sometimes, however, they can be seen to divide into three or more filaments before so doing.

Fig. 4, c.—The pneumogastric, before reaching the aorta, gives off a branch, which, winding around the arch, changes its course upwards, and soon divides into two nerves; one crossing the neck enters the œsophageal tissue—the other, the recurrent laryngeal (Fig. 4, c), joins the trachea, and, in close contact with its side, follows it to the larynx, and enters the crico-arytenoid muscle. There are no fibres from the recurrent distributed to the crico-hyoid directly, or indirectly through communication with the superior laryngeal.

CHAPTER II.

THE preceding chapter has been altogether taken up with anatomical descriptions of the respiratory organs and their appendages. So much that was new was met with during our dissections, that it was thought better to separate the description of the anatomy from the physiological statements. We have thus the physiology of the respiratory organs still to describe, and this can now be done without repeating any more of the anatomical detail than is necessary to enable the reader to comprehend the actions of the organs concerned.

The history of the theories entertained as to the nature of the respiratory motions in turtles, appears to us one of the most extraordinary in the records of science. Totally misunderstood by the earlier naturalists and biologists, or confounded as to type with the respiration of Batrachians, this function in turtles was first rightly comprehended, at least to some extent, by R. Townson in the latter part of the last century. How far he went, and how far he was correct, we shall more fully point out in another place. The authority of more eminent naturalists, and an obstinate disposition to associate the turtle with the frog, and to insist on similarity as to the execution of their functions, gradually drew attention from Townson's statement, and more modern authors have paid it no deference whatever; yet, as we shall distinctly show, all the later writers are utterly wrong, and his opinions as to the facts in question are thus far the only ones which seem to be correct. In reading his very ingenious essay, which we have elsewhere quoted at length, p. 2, it is hard to see how the statements and evidence could have failed of more respectful and permanent attention. A complete review of the theories entertained in regard to the respiratory function in Chelonian reptiles, will more fully illustrate the above remarks.

As early as 1719, Malpighi[1] described the respiration of turtles as similar to that of frogs. Both alike were supposed to distend the lungs by swallowing air, so that, in place of air being drawn into the lung-sacs, it was forced into them by the movements of parts above the trachea; but while in the frog this was effected plainly through the aid of the bellows-like mouth, in the turtles their vast hyoid apparatus was by some supposed to constitute a forcing pump of similar purpose and nature. The authors of Malpighi's era shared these opinions, and with the one notable exception above mentioned, they have stood almost uncontradicted up to the date of a paper by one of the authors of the present essay.

---

[1] Adversaria Anatomica, t. v Animadv, 29.

The latest and best work on comparative anatomy and physiology[1] thus describes its author's conclusions as to this subject: "C'est aussi par des mouvements de déglutition que la majeure partie de l'air inspiré est poussée dans les poumons chez les tortues; mais ici ce mode de respiration est nécessité par une disposition organique inverse de celle que je viens de signaler chez les Batraciens." M. Edwards then proceeds to point out the rigid form of the turtle's frame, the absence of mobile ribs, and the consequent necessity for the belief that the lungs in these animals cannot be dilated from without, as occurs in mammals. The same opinion is held by nearly all writers at the present time; but some, in place of describing the process as one of deglutition, effected alone by muscles on the floor of the mouth, regard the hyoid apparatus as the true forcing pump concerned in propelling air into the interior. Thus, T. Rymer Jones,[2] after describing the fixity of the bones of the chest in turtles, adds, that "under these circumstances, as a compensation for the want of mobility in the chest, the os hyoides and the muscles of the throat are converted into a kind of bellows, by which the air is forced mechanically into the lungs, and they are thus distended at pleasure." In fact, the submaxillary space with the hyoid arches, are in continual motion in turtles, and this movement precisely resembles the like action in frogs; but while in these latter it is really a respiratory act, in turtles, as we shall show, it has other purposes, and, while it has deceived observers, may be proved to have no influence of any moment in carrying on the breathing process. Muller[3] gives a like account, and adds, that expiration is effected by means of muscles between the lower shield or plastron, and the posterior extremities. Carpenter[4] has a brief description of the respiration in chelonia, which corresponds to the general opinion already quoted above.

Prof. Agassiz's description[5] being one of the latest, and certainly one of the most authoritative statements, we quote in full, to complete our history of the generally received ideas as to the mechanism of chelonian respiration.

"Here, again, we meet with a very striking ordinal character. The turtles swallow the air they breathe. The breast box, which includes the lungs, being immovable, a respiration like that of the other reptiles, the birds, and mammalia, performed by the expansion and compression of the breast box, and consequently of the lungs, is impossible. Owing to the peculiar structure of their trunk, breathing is therefore only possible for turtles, by a pressure of the air from the mouth down into the lungs; but though we are persuaded that this swallowing of the air constitutes the main act in the process of breathing, still we are inclined to believe, against the opinion of other anatomists, that the diaphragm, which in turtles is very much developed, and attached to the lungs, takes also its part in that act. Moreover, the muscles of the shoulder and of the pelvic region may assist in that

---

[1] Milne Edwards, Leçons sur la Physiologie et l'Anatomie comparée de l'Homme et des Animaux, t. deuxième, deuxième partie, p. 387. 1858. Paris.

[2] The General Structure of the Animal Kingdom, p. 567.

[3] Physiology—London translation, p. 360, vol. ii.

[4] Gen. and Comp. Phys., p. 493.

[5] Contributions to the Natural History of the United States, vol. i. p. 281.

operation, either by immediately compressing the lungs, which generally extend in turtles from one end of the trunk to the other, or by pressing the bowels against them.

" The act of swallowing the air is chiefly performed by the apparatus of the tongue-bone, and the tongue itself, which, by its large size, facilitates the operation. Being drawn backwards and upwards, this organ shuts up the choannæ, and at the same time opens the slit of the windpipe, situated just at its base, thus giving to the air a passage into the windpipe, and at the same time preventing its entrance through the choannæ into the nose. In this way, the tongue takes the place, in a certain sense, of the velum palatinum of the higher vertebrata, which is wanting in turtles. After the air has passed into the windpipe, the tongue is drawn forwards, and thus the longitudinal glottis is again closed, while now the choannæ are again opened to a free communication with the cavity of the mouth."

Professor Agassiz adds, in a following note:—

" We find the same mode of breathing in the class of Batrachians, but for an entirely different reason, namely, on account of the absence of ribs."

Also.   " The existence of a diaphragm is erroneously denied to turtles by Dumeril and Bibron, Erpétologie générale, 1, p. 175."

In the above description, Prof. Agassiz exhibits some doubt as to the correctness of received views on this subject, and speaks of the musculus diaphragmaticus (Bojanus) as having something to do with the act of respiration, which he thinks may also be aided by other muscular parts, as those concerned in locomotion, and by certain pelvic muscles which he does not specify by name.

We shall show as we proceed that, although the muscle covering the lungs may be homologous with the diaphragm of mammals, it is really a muscle of *expiration*, and therefore not analogous to the diaphragm when regarded from a physiological stand-point.

Except for the purpose of completing this brief history of opinions held now or abandoned, it is only requisite to allude to the views of Perault, who attributed the inspiratory act to the elasticity of the lungs, and the expiratory motion to muscles of which, he says naively, the turtle has an abundance. M. Tauvry, whose views Milne Edwards partially indorses, attributed the whole respiratory act to the changes in the capacity of the chest, caused during locomotion, by the advance of the head and limbs from and their retraction within the carapace. M. Haro[1] supports the same views, but, although both are successful in showing that these movements may alter the capacity of the chest-box, and thus under some circumstances modify respiration, neither has proved that respiration relies for its continued occurrence upon these motions, nor would such a supposition be entertained for a moment by any one who surveyed the mechanical conditions which are effective in carrying on respiration in other animals. That the locomotive movements may, and perhaps do at times *modify* the respiratory process, may be taken for granted. That other agents are constantly employed in this function is not less clear, nor shall we have any difficulty in disproving M. Haro's theory by unanswerable facts.

---

[1] Mém. sur le respiration des Grenouilles, Ann. des Sc. Nat.   2 serie, t. xviii. p. 48.

The author to whom we have alluded as the only one who has approached to a clear comprehension of the true mechanism of respiration in turtles is Robert Townson, LL. D.[1] The anatomy of the respiratory muscles of the breast-box is described by this author, as we have elsewhere shown, with much correctness. His statement as to the mechanism of the movements of the chest and belly muscles in breathing are, also, remarkably truthful, and are approached in this particular by those of no other or later authors.

He came to the conclusion, as we have seen, p. 6, that the turtle and frog do not breathe alike, but that while the latter forces air into the lungs, the former possesses a type of respiratory movement closely analogous to that of the mammal.

He described an inspiratory muscle in the posterior flanks, and an expiratory muscle covering the back of each lung, and attached to a broad tendinous expansion, running forward, to be inserted in front on the carapace, above the lung. To do full justice to this most ingenious and neglected observer, we have quoted, in connection with the anatomy of our subject, the experiments, by means of which he proved that turtles do not force air into the lungs, p. 6, and by which he also showed that they draw the air into the chest, by muscles attached to the breast-box, and expel it through the aid of the expiratory muscle covering the posterior end of the lung.

Considering the period at which he wrote, nothing could be clearer than the above statement, and we are amazed, that its obvious truth should have so long escaped recognition.

In the summer of 1861, one of us, Dr. Weir Mitchell, while engaged in studying the blood-pressure in the snapping turtle, Chelydra serpentina, became convinced that the prevailing views as to the respiratory mechanism of Chelonian reptiles were totally incorrect. Accordingly he partially studied the subject, and incidentally embodied his opinions in an essay upon the blood-pressure in the snapping turtle.[2] At the time referred to, Dr. Mitchell was unacquainted with Townson's researches. The views of Dr. Mitchell, and the experiments by which he supported them, will be found scattered through the text of the present essay, of which, indeed, they form the basis. In the summer of 1862, the present authors took up anew the study of the respiration in turtles, and have endeavored to render it as complete as possible. In so doing they have been fortunate enough to carry the subject far beyond the crude experiments of Townson, and to discover anatomical and physiological facts of the utmost interest and novelty, which have hitherto escaped attention.

To facilitate the comprehension of the subject, we shall divide the physiological part of this essay in the following manner:—

1st. The externally visible phenomena of respiration.
2d. Physiology of the muscles of respiration.
3d. Physiology of the respiratory nerves.

---

[1] Tracts and Observations in Natural History in Physiology. London, 1799. Cuvier's views and his criticism of Townson may be found appended to the full quotation of Townson's dissertation, at p. 6 of this essay.

[2] American Phil. Trans., Phil. 1862.

When a turtle of any kind is observed with care, it will be seen that it breathes at very irregular intervals. These are much prolonged when it is in the water, and half an hour or more may elapse before it rises to the top, to take two or three respirations, preparatory to a second plunge. When, during summer weather, the snapping turtle was placed on a table, and observed in air, its respiration averaged one to every two minutes and a half, although certain individuals breathed more rarely, and all irregularly. The box turtle breathes still less frequently. A large snapper observed for some time, gave the following record:—Ten respirations were noted with the intervals between them, which were as follows:—1, 2, 1, $\frac{1}{2}$, 5, $\frac{1}{2}$, 1, 4, 3, 2, $\frac{1}{2}$ minutes respectively. In another the respirations during an hour were at almost perfectly regular intervals of two minutes. The size of the turtles did not seem to bear any notable relation to the number of respirations per minute.

During the respiratory act in the snapping turtle, C. serpentina, the box turtle Cistudo Virginea, the green turtle Chel. mydas, and several Emydæ, we have noticed carefully the exact details of the motions of the various parts. The head and neck, the flank spaces in front of and behind the limbs, these themselves, and the mouth, glottis, and hyoid apparatus, have been scrutinized with care in hundreds of instances, and with these results.

Turtles breathe easily with the mouth open or shut. This fact alone deprives their respiration of all resemblance to that of Batrachians.

The respiratory process is threefold, and consists of—

1. Complete expiration.
2. Complete and very full inspiration.
3. An appearance of slight, or partial expiration, followed by a pause of greater or less duration.

During the period which precedes this series of movements, the turtle being at rest, the spaces between the posterior members and the plastron and carapace are nearly level, or only a little concave. The shoulders are pushed forward somewhat, the lungs being full at this time, while the large hyoid apparatus is usually dilated or drawn backwards and downwards. Sometimes it is in continual motion, like that of the frog when breathing, but in the turtle this rise and fall of the hyoid arches has no essential connection with that function. When, during the inter-respiratory pause, we open the jaws the same movements of the hyoid apparatus may still be seen, nor is it easy at these times to assign to them any very obvious purpose. The glottis may be seen at rest, as a linear slit, Fig. 7, A, in the centre of an ovoidal slightly elevated mound, just back of the tongue, on the floor of the mouth. The first respiratory act is one of expiration. Whether the mouth be opened for observation or not, the following movements occur: The hyoid apparatus descends and broadens laterally especially at its posterior part, carrying the glottis back and a little down. The object of this action we suppose to be, the separation of the glottis from contact with the roof of the mouth, in order that the air may the more readily enter it after passing through the nares. At the moment of beginning to expire the glottis opens wide, so as to form a rhombic figure (Fig. 7, B.) It remains thus until the whole respiratory act is completed. Meanwhile, during expiration the limbs fall in towards the shell quite passively, and the flank spaces in front of the posterior limbs sink so as to present deeply concave surfaces.

A   Fig. 7.   B

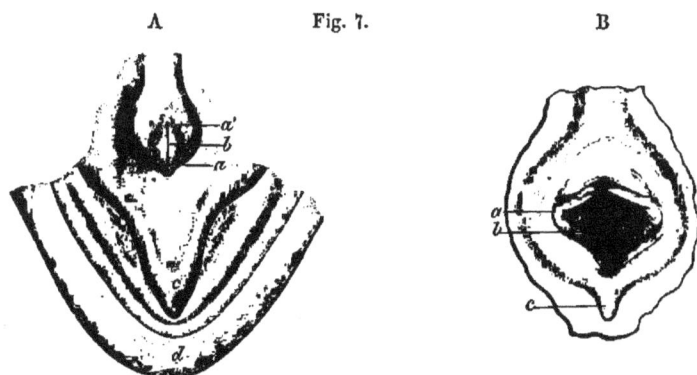

Fig. 7, A. The glottis closed.—*a a'*, the line formed by the glottis lips when the animal is not breathing; *b*, the prominent central part of the glottic lips, indicating the summit of the arytenoid cartilage; *c*, tongue; *d*, lower jaw.

Fig. 7, B. The appearance of the glottis during respiration.—*a*, right glottic lip; *b*, rima glottidis; *c*, extremity of the hyoid bone.

A full inspiration instantly follows. The flank spaces become flat and tense, rising to a level. The glottis remains open. The hyoid arches advance, and at the close of the inspiration the shoulders are pushed passively forward.

As soon as the lungs are completely filled, a very slight expiration relieves them of the surplus air, the flank spaces sinking a little, the hyoid arch at rest, the glottis closing at the end of the expiration. The final action here described appears to be due to the cessation of activity on the part of the inspiratory muscles and to the passive falling in of the limbs displaced during their contraction. The lungs are thus left full of air, and ready for the next act of respiration. Whenever a turtle in air breathes, these triple actions occur, but when under water it occasionally expires air, and does not rise to renew the supply until some time has passed by.

Type of respiration in Chelonia.—We are now prepared to examine the subject from another point of view. A superficial observer, or one who accepts the present belief, sees in the motions of the hyoid arches a movement in appearance corresponding to the respiratory play of the floor of the frog's mouth. Yet the slightest anatomical examination should have shown that, while in the frog the nostrils have valves essential to their mode of breathing, in the turtle there are none, while the form of the horny lips in the latter animal renders it impossible to make the mouth so air-tight as to act the part of a chamber in the supposed process of pumping air into the lungs. On the other hand, the laryngeal cavity is also too small to act as a chamber, nor does the hyoid arch, in its descent, enlarge the laryngeal area.

When, at the beginning of this research, one of us observed the turtle (snapper) breathing *with an open mouth*, while watching a chance to bite, he was at once convinced that the agents of respiratory movement were below the trachea, and the

4

following very simple experiments converted this conviction into the most absolute certainty—a certainty which every future step served but to illustrate from new points of view.

On page 77 of the memoir of Dr. Weir Mitchell, previously cited, are to be found the experiments above alluded to. The trachea of a large snapping turtle was cut across, after which breathing went on at the usual rate, or more often, owing to causes presently to be mentioned. Next, a bent glass tube, two millimetres in width, was adapted to the upper or outer end of the divided trachea, and allowed to dip into water. If the breathing power resided in the hyoid arches, larynx, and mouth, the water in the tube should have been forced downwards during inspiration, but, although respiration continued, the fluid moved at this time only about one millimetre, and even this was plainly due to the motion of opening and closing the glottic lips, which occurs synchronously with the respiratory movements in the breast-box.

The same bent tube was next adapted securely to the lower end of the divided trachea, and again dipped into water as before. At each subsequent inspiration the water was forcibly and largely drawn up into the lung, and again rejected during expiration. After this no doubt could exist as to the locality in which arose the mechanical force productive of respiration. With this convincing proof the subject was laid aside for the future and more thorough investigation, of which this essay is the record.

Function of the respiratory muscles of the Turtle.—A large snapping turtle was secured on its back, and an incision made over the flank space, between the posterior limb and the plastron and carapace. The skin and superficial fascia were then carefully removed so as to expose the whole muscle which fills this space, and which has already been fully described.

When inspiration took place, the muscle contracted, and as it is possessed of a central tendon from which radiate fibres in all directions, the result of their shortening was to convert its previous deeply concave surface into one which was nearly level, while at the same time the air rushed through the open glottis into the lung. The analogy between this muscle and the diaphragm of mammals was absolutely perfect. The central tendon, the converging muscular fibres, and the form of movement resulting from this beautiful arrangement, all united to suggest the resemblance. The inspiratory function of this muscle was palpably evident, nor could any other office be possibly assigned to it, because it was attached to no movable bone or other parts susceptible of motion.

Repeated galvanization of this muscle served further to demonstrate its purpose. Finally, the muscles on both sides were removed, when all inspiratory power was lost. The turtle could empty its lungs, but possessed no power to fill them anew.

The muscles engaged in expiration were next made the subject of study. At first we were led to believe, that the elastic contractility of the lungs might alone suffice to empty them, but this was opposed to all physiological analogy, and the power with which expiration occurred was too great to allow us to suppose that no muscular force intervened for its production.

To examine this part of the subject, a turtle (snapper) was secured, as usual, and the plastron removed, with the exception of a rim at the back and on each side, to which remained attached the fibres of the inspiratory muscles. After a few minutes the turtle expired the air in the lung. During this action, the fascia covering the lungs below, and lying between the peritoneum and the plastron, was observed to become tense, owing to the contraction of the two sheets of muscle, which terminate this tendon anteriorly and posteriorly, and find origin in the carapace.

Recalling the full anatomical description already given, it will be remembered, that the lungs and abdominal viscera are covered outside of, and below the peritoneal sac, by a white membranous tendon, which extends across the middle line, and is firmly attached to the pericardium, as well as by firm areolar tissue to the central line of the plastron or lower shell. The muscular bellies arising from this covering tendon, fold over the lung in front and behind. Opposite to the inspiratory muscles are also areolar fibres, binding its tendon to the fascia of the expiratory muscle above it. When the four bellies of this muscle, or muscles contract, the lungs are acted upon directly, or by being compressed through the medium of the other viscera which are, so to speak, grasped during this powerful movement. At the same time, the passive inspiratory muscles are drawn up with the retreating lungs, owing to the pressure of the external air, and to the close union between the two sets of antagonistic muscles. Although the pericardium is also fastened to the expiratory tendon, this sac is so firmly bound to the plastron below it, that it does not appear to be disturbed during expiration, unless the connecting fibres are divided, in which case the heart sac and its contents are strongly drawn from the plastron, as the air is expired from the lung.

As in the case of the inspiratory muscle, the expiratory muscle was also tested by observing its action when exposed in the living animal, and by galvanizing its fibres. The purpose of this singular sheet of muscle and connecting tendon admits then of no doubt. Aided by the elasticity of the lung, it empties that viscus of air, and no other muscle appears to lend it any aid.

The third period of respiratory movement is marked by the closure of the glottis, and by the relaxation of the muscle of inspiration, the limbs then settling passively to their new positions. Hence the general appearance of a slight expiration at the end of the inspiratory act.

It is impossible to review this account of the respiration in chelonians, without being struck with the simplicity of the plan. A box containing all the viscera of the chest and belly has an open space on each side, filled by a muscle of peculiar form, whose contraction increases the size of the visceral cavity, and thus causes air to rush into it. Within the breast-box, the lungs and visceral mass embraced by a single muscle, obey its contraction in effecting expiration, and as the visceral cavity thus becomes smaller, the inspiratory flank muscles curve in to fill the gap.

After the most careful investigation, we can discover no other respiratory muscles within the breast-box.

The muscular apparatus of the glottis is equally simple. There is a muscle to open it, and another muscle to close it. Here, as in the rest of this portion of our essay, we shall not commit ourselves by names, which, although they may recognize homologies, confuse the reader, who has sometimes to bear in mind that their functions may be exactly the reverse of those of the human muscle whose name they carry.

The two glottic muscles have already been fully described; when both are cut away or paralyzed, by section of their nerves, the glottis still closes, owing to the elasticity of its cartilages, but it does not shut firmly, and if the lungs be previously filled with air, a large part always escapes. Under ordinary circumstances, the glottic lips are closely pressed together by the sphincter-like muscle which we have described and figured. The mass of its fibres lie below the opening muscle, and are parallel to the direction of the glottic lip, while its connections are principally at the anterior and posterior end of the glottic line. When contracted, as it always is more or less strongly during the interval between two respirations, it would tend to pucker the glottis somewhat, if it were not that the anterior and posterior insertion are firmly fixed, by the parts in front of and behind them respectively. Thus attached, the only influence it can exert, is to close the glottis whose lips stiffened by the arytenoid cartilages facilitate the process.

The opening muscle lies outside of the closing muscle, nearly at right angles to it, and immediately under the mucous membrane of the glottic mound. At the moment when expiration begins the respiratory act, this opening muscle contracts so as to draw the glottic lips wide open and permit the air to escape. Then follows a full inspiration, the glottis still open, and lastly it is closed by the constrictor muscle just after the great flank muscles of inspiration cease to act.

The downward movement of the hyoid arches is effected by the omo-hyoid and other muscles of the neck. It appears to be intended to remove the glottis from contact with the roof of the mouth during the act of respiration. The upward motion of the hyoid apparatus is produced by a thin sheet of muscular fibres spread transversely across it and over the whole upper part of the neck.

The function of all of the above muscles was determined by simple observation, by stimulating them directly, and by irritating their nerves.

The necessity for closing the glottis firmly in these animals becomes obvious, when we reflect, that not only must they be enabled to retain the air, but when under water be competent to exclude that fluid from the lungs. In fact, when we divide the trachea, or in any way paralyze the glottic muscles, the power of retaining air in the lungs is totally lost for a time. The moment the respiratory muscles cease to act, the elasticity of the lung asserts itself, and that viscus is immediately emptied. After a day or two, however, a curious change may be noticed; the turtle breathes as usual, but in place of allowing the air to escape through the open trachea, the animal holds the inspiratory muscle contracted, and thus retains the air in the lung a considerable time after each inspiration. There seems to be some urgent necessity for thus holding the air a long time in the lung, and perhaps for keeping the lung distended. The instinctive provision for these purposes when the usual means fail, is well worthy of note. As we proceed with the study of the laryngeal nerves, we shall have further occasion to observe the great importance of the glottis, and to wonder at the singular means to which creative power has resorted, in order to secure the orifice from the ordinary chances of accident and disease.

The physiology of the nerves of respiration in turtles has been the subject of our most careful and complete study. So novel and surprising were some of its results that we have felt it right to surround ourselves with more than common precautions. For this purpose we have repeated our experiments and dissections on several species of turtles, and on numerous individuals of each species, until incessant repetition left no question unanswered, and no conclusion doubtful.

We shall study,

1st. The physiology of the pneumogastric nerve and its branches, so far as they concern the respiratory function.

2d. The physiology of the nerves which supply the respiratory muscles of the breast-box. For all necessary details as to the anatomy of the vagus nerve and its branches we refer to the former part of this memoir. Here it will only be requisite to repeat that, as in most mammals, the larynx receives a superior laryngeal nerve, and an inferior or recurrent laryngeal trunk. The superior, which in man is the nerve of sensation to the larynx, is in turtles distributed to the mucous membrane of that organ, and also to *both* of the glottic muscles. The recurrent laryngeal, which in man is the principal motor nerve to the larynx and glottis, is in turtles also motor, but it sends branches only to the opening muscle. The remaining peculiarities will be better understood as we proceed to state in sequence the experiments which led to their discovery.

Experiment.—A large turtle (snapper) was secured on its back, its mouth held open. It breathed well at intervals of two minutes or more. The recurrent nerves were exposed and galvanized at the middle third of the trachea. Irritation by this agent and by mechanical means, caused the lips of the glottis to open, although not very freely. The two nerves were then divided, and the trachea cut across. The glottic movements continued perfect, and were synchronous with the respiratory motions of the breast-box. The muscles of the right side over the hyoid apparatus were then removed, the covering fascia beneath them dissected off, and the superior laryngeal nerve discovered lying under the shelter of the superior hyoid wing. Irritation of this nerve or its fellow on the opposite side caused the outer edge of the glottic lips to open, while the inner edge appeared to be forcibly closed at the same time. On cutting the nerves across, and stimulating the peripheral ends, like results were observed.

The left superior laryngeal nerve being intact, galvanization of the centric end of the divided nerve on the right side caused first, closure of the inner lips and opening of the outer lips of the glottis; and second, violent and general muscular movements and winking, apparently expressive of acute pain.

Finally the left superior laryngeal nerve was divided, when complete paralysis of the glottis ensued.

Order of section, and results:—

1. Section of both inferior laryngeal nerves, causing glottis to open; glottic movements perfect after section.

2. Cut right superior laryngeal nerve, causing glottis to open superficially and to close below; galvanization of outer end of nerve caused same result; galvanization of centric end gave signs of sensibility and reflex closure of glottis, and opening of its outer lips.

3. Section of left superior laryngeal nerve; complete paralysis of glottis.

Experiment.—A small snapper was secured as usual, and the hyoid apparatus separated from the lower jaw and turned up for convenience of observing glottis. We then cut subcutaneously the left superior laryngeal nerve, causing motion in the glottic lips. This section slightly lessened the power to move the glottic lips on the side cut. We next divided, in like manner, the right superior laryngeal nerve. The power to open the glottis remained but little impaired, but the air could no longer be retained in the lungs. Respiration went on as usual, but when inspiration was complete and the muscles relaxed, the glottic lips fell together by virtue of their own elasticity, although this seemed insufficient to balance the contractile force of the expanded lung, whose contents therefore escaped. Then followed renewed inspiratory efforts, necessitated by the loss of power to close the glottis, until the animal learned to hold the air in its lungs by keeping tense, for a time, the flank muscles of inspiration. The left and right inferior laryngeal nerves having next been divided, entire paralysis of the glottis ensued, the flaccid lips falling together valve-like when efforts were made to inhale air, while, if air was blown into the lungs, it escaped without difficulty.

Order of section, and results:—

Section of left superior laryngeal. { Glottic lips convulsed by section.

Section of right superior laryngeal. { Loss of power to close glottis firmly.

Section of both recurrent laryngeals. { Complete paralysis of glottis; loss of power to open glottis.

The above experiments, repeated upwards of twelve times on the Chrysemys picta, the Cistudo virginea, the Chelonia mydas, and the Chelydra serpentina, left no doubt in our minds as to the functions of the two laryngeal nerves in turtles. Careful dissections enabled us moreover to trace these nerves so as to show that, while the inferior laryngeal is distributed only to the opening muscle of the larynx, the superior laryngeal sends branches to both the dilating and the constricting muscles.

This anatomical arrangement explained to us some of the difficulties which we had encountered while testing the function of the muscles by means of irritants applied to the nerves. Thus, when the upper nerves were irritated, the glottis opened at the outer lip and closed within, because the irritant necessarily acted both on the nerve fibres of the closing and of the opening muscles. Again, when the lower nerve, inferior laryngeal, was galvanized, it caused the lips of the glottis to open, but not freely, because the motion of the lips seemed to act reflectively as a cause of irritation through the mucous branches of the superior laryngeal on to its nerve centres, and thence by its motor fibres upon the opponent closing muscles. When, however, the superior laryngeal nerves were cut, the closing power was abolished, and then, irritation of the inferior nerves produced more perfect dilatation of the glottic chink. We have thus determined by every necessary means that the superior laryngeal nerves in turtles are the nerves of sensibility for the mucous membrane of the larynx and glottis. That they are the motor nerves of all the true glottic muscles, and enjoy thus the ability to open and to close this orifice, and that the inferior laryngeal nerves are the motor nerves of the dilating muscles only, and have not sensibility or power to close the glottis.

What then is the reason of this double distribution of two nerves to one muscle? Upon this question we shall presently return. It seems highly probable that both nerves usually act at once to open the glottis, since galvanization of either set of nerves does not fully effect this end, while, when both sets of nerves are stimulated, the glottis opens wide.

The distribution and functions of the two laryngeal nerves in turtles are thus seen to be totally different from what we see in mammals. In them, as we need only to remind the reader, the superior laryngeal is a nerve of sensation chiefly, and although it possesses also a motor filament, this, in man at least, is distributed to a muscle, the crico-thyroid, which has neither homologue nor analogue in chelonian reptiles. In mammals the inferior laryngeal is, as in the turtle, a motor nerve, but it supplies alike the dilating and the closing muscles of the glottis.

On reference to the anatomical part of this essay, it will be seen that the hypo-glossal nerve lies close to the track of the superior laryngeal nerve, and might readily be confounded with it, when the intention is to find and divide the latter alone. The nerve in question supplies muscular branches to the tongue only.

Thus far the physiology of the glottic nerves in turtles, although determined for the first time, and shown to present points of great interest and novelty, has not exhibited any peculiarity so exceptional as that to which we shall now direct attention.

This was brought to our notice while further pursuing the study of the functions of the glottic nerves. The mode in which it was first suspected, then discovered, and finally set in clear light by every available means, will be best set forth in the following record of our experiments and inferences, in the order in which they occurred.

Experiment.—A small snapper, one and a half pounds in weight, was secured as usual. Its respiratory acts observed to be perfect, and the two inferior laryngeal nerves divided one after the other, causing twitching of the glottic lips. After this the glottis still opened and shut as before, and, indeed, equally as well. It was plain, as we have already seen, that the superior laryngeal nerves could open and shut the glottis without other aid. Next, the right superior laryngeal nerve was cut at the middle of the upper hyoid cornu, and the glottis was carefully observed.

The section caused twitching of the glottic lip, and at the next respiration, to our great surprise, *both* sides of the glottis, the right as well as the left, opened equally well. In fact there was no difference. A close inspection satisfied us that the section of the nerve was complete.

If now we recall the facts, that the glottis of both sides was moving despite the section of both recurrents and one superior laryngeal nerve, it will be seen how mysterious this must have appeared to those who first observed it. We came to the conclusion either that there existed some mechanical arrangement of the glottis and its muscles, which enabled one side, while in motion, to communicate that movement to the other, or, that there was a direct nerve communication between the right and left superior nerves of the larynx. The first hypothesis was unsupported by anything that we knew of the parts. The second seemed unlikely, since on reflection we could recall no instance of a true chiasm of any nerves except those of sight. We hastened to examine the question by new experiments.

Experiment.—Snapper, weight two pounds. We exposed and galvanized the left inferior laryngeal nerve, thus causing both lips of the glottis to open. The same result was obtained with the right nerve. This fact, observed by us in other cases, was soon found to be due to the difficulty of insulating the current in one nerve. When, however, we made use of mechanical irritants, stimulation of one nerve affected only the glottic lips of the same side.

The right inferior laryngeal nerve was then cut, and immediately afterwards the right superior laryngeal nerve. The glottis still moved as well as before these sections. Next, we cut the left recurrent (inferior laryngeal nerve), thus leaving the left superior laryngeal the only nerve entire. Nevertheless, the glottic lips on both sides opened and shut, as well and as completely as ever. Lastly, we cut this remaining nerve, causing total paralysis of the glottis, and the usual results as to respiration.

Order of section, and results :—

1st. Cut right recurrent nerve (inferior laryngeal) and rig t superior laryngeal nerve; glottis continues to move perfectly on both sides.

2d. Cut left recurrent (left inferior laryngeal); glottic action perfect on both sides.

3d. Cut left superior laryngeal nerve ; total paralysis of glottis.

Experiment.—Snapping turtle, weight three and a half pounds. We dissected the hyoid apparatus from its connection with the lower jaw, and held it back, thus freely exposing to view the chink of the glottis. Up to this time we had reached the conclusion, that somewhere on the fenestrum in the cricoid cartilage there might be a branch of communication between the two superior laryngeal nerves of the larynx. Therefore, on the turtle prepared as above described, we made an incision on to the fenestral membrane, between the larynx and the hyoid bone, opposite to the junction of the superior cornu with this bone. The section made a little to the left of the median line caused slight twitching in the glottic muscles, but had no influence on the respiratory motions of the glottis.

The two inferior laryngeal nerves were next divided, and still the glottis moved as perfectly as before. The left superior laryngeal nerve was divided at the middle of the upper hyoid cornu, and immediately all motion of the left side of the glottis ceased, the right side moving during respiration as usual, although somewhat feebly, owing perhaps to loss of blood during the first part of the experiment.

Section of the right superior laryngeal nerve completed the paralysis of the glottis.

Order of section, and results :—

1st. Section through supposed site of communicating nerve; no effect as to respiratory movements.

2d. Section of both inferior laryngeal nerves ; no further effect of any permanent nature.

3d. Section of left superior laryngeal nerve ; paralysis of left glottic lip.

4th. Section of right superior laryngeal nerve; complete paralysis of glottis

The above experiments led us, irresistibly, to the conclusion, that there must be a chiasm of the two superior laryngeal nerves, and it only remained to prove, with the scalpel, the presence of this branch. A careful series of dissections on large turtles of various species and genera, satisfactorily proved that we were not mistaken. In every case the nerve was readily found, and the physiological prediction as to its existence verified in the most absolute manner.

The discovery of a new nerve in turtles, and upon ground over which the accurate knife of Bojanus had passed, called for a still more rigorous testing of our previous results. For this purpose the following experiments were made.

The first of this second series is of unusual value, owing to circumstances which arose incidentally.

Experiment.—Snapping turtle, weight nineteen and three-quarter pounds. We cut down on the middle line of the hyoid bone and divided it throughout its length with a hair-saw and nippers. When this operation is done with care, it exposes to the operator enough of the cricoid fenestrum to enable him to cut the communicating nerve at its central part. Next, both recurrent nerves were divided at the middle of their course. The section, and after stimulation of the right nerve, had no effect on the glottis, which we thought singular. Section of the right superior laryngeal nerve was satisfactorily made as usual, the nerve being readily exposed and divided. To our surprise, the right glottic lip became paralyzed almost totally, the left side moving in respiration as usual. This result was opposed to all our former experiments. After a rigid examination of the conditions of this last experiment, and finding in them no explanation of the contradiction which it offered, we dissected, with scrupulous care, the whole track of the pneumogastric nerve and its branches to the larynx, as well as that organ itself. The following appearances were noted: On the left mucous lip of the glottis, a small white patch of diseased tissue. The inner end of the right upper hyoid cornu was enlarged to double its normal size; thus of necessity stretching the right superior laryngeal nerve where it crosses the cornu at its inner end. On the left side the superior laryngeal nerve was perfect up to the point at which it gave off the interlateral communicating branch. This latter nerve, lying on the cricoid fenestrum, was involved in a mass of diseased tissue, which extended between the trachea and the body of the hyoid bone, from its lower part to a point about one-quarter of an inch above the fenestrum. This disease, doubtless, affected the communicating branch, so as to cause partial paralysis of the right glottic lip to follow section of the corresponding superior laryngeal nerve. Had the interlateral branch been completely destroyed, section of one laryngeal nerve must have produced *entire* paralysis of the glottic lip on the side operated upon.

This observation, which at first promised to cast doubt upon those which preceded it, thus proved at last the most conclusive evidence of the correctness of the view to which we had arrived. An accident of disease or injury had so altered the communication between the two superior nerves of the larynx, as to make unnecessary the section, which would under ordinary circumstances have followed as the third step in the experiment.

5

Experiment.—This experiment was designed to be a repetition of the plan of the last one, but in dividing the hyoid bone to reach the nerve at the middle line, the saw, accidentally carried too deep, touched the membrane on which runs the nerve. Section of the recurrents followed with the usual negative result. Section of the right superior laryngeal nerve produced paralysis in the right glottic lip. If our former view be correct, then in the present case we must have cut the communicating branch with the saw. In the above experiments, the sections and results may be thus stated:—

1. Section of interlateral communication between the two superior laryngeal nerves; glottic respiratory motions as usual.

2. Section of both inferior laryngeal nerves; glottic respiratory motions as usual.

3. Section of right superior laryngeal nerve; paralysis of right lip of glottis.

Experiment.—Snapper, weight four pounds. We cut first the two inferior laryngeals; next we divided the right superior laryngeal. The glottic movements were still perfect. One nerve was sustaining unimpaired the whole ordinary motions of the glottis in respiration. Indeed, the closest scrutiny failed to discover in its action any departure from the condition of health. Lastly, we sawed through the hyoid bone, glottic acts still regular. Then with a hook we lifted the nerve and divided it. Instantly a respiration followed, but the right glottic lip was now motionless.

Order of section, and results:—

1. Section of both inferior laryngeal nerves.

2. Section of right superior laryngeal nerve; after which the glottis moved in respiration as usual.

3. Section of median intercommunicating nerve; paralysis of right glottic lip.

Experiment.—This turtle had been used for other purposes, and had undergone an hour before section of the middle cervical spine. The respiratory motions of the breast-box had ceased, but at intervals the glottis opened and closed with normal regularity. The trachea was divided, and with it both recurrent laryngeal nerves. Next we cut the interlateral communicating nerve. The glottic acts still remained perfect. Lastly, we exposed the left superior laryngeal nerve, and divided it, causing instant paralysis of the left glottic lip.

Order of section, and results:—

1. Section of both recurrent laryngeal nerves.

2. Section of communicating branch; glottic acts perfect.

3. Section of left superior laryngeal nerve; paralysis of left glottic lip.

As further illustration, we give in brief the order of section and results in two box-turtles.

Experiment.

1. Section of both inferior laryngeal nerves; glottic motion perfect.

2. Section of right superior laryngeal nerve; glottic motion perfect.

3. Section of communicating nerve; paralysis of right lip of glottis.

4. Section of left superior laryngeal nerve; total paralysis of glottis.

Experiment.

1. Section of communicating nerve.

2. Section of right superior laryngeal; glottic acts perfect, perhaps not closing

firmly on the right side; the right glottic lip now relied alone on the recurrent nerve for opening power.

3. Section of right recurrent (inferior laryngeal nerve); paralysis of right glottic lip.

The above stated experiments were repeated very frequently, and always with the like results. If the evidence which we have given be reliable, we have now proved that in turtles there exists a communication between the right and left superior laryngeal nerves, of the nature of a true *chiasm* precisely like that of the optic nerves, and, so far as we know, the only instance thus far discovered of this anatomical peculiarity in nerves exterior to the great centres.

Fig. 8.

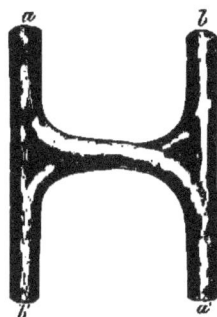

Fig. 8. Diagram of the chiasm of the superior laryngeal nerves.—*a a'*, intercommunicating fibres of the right nerve; *b b'*, similar fibres from the left nerve.

The diagram, Fig. 8, illustrates our views in regard to the track of the nerve fibres. Part of each nerve probably proceeds directly to the two glottic muscles of its own side, while another strand crosses over through the interlateral trunk to be similarly distributed to the two muscles of the opposite side.

Keeping this in view, we can now see how one single superior laryngeal nerve may move the glottis on both sides, until the chiasm is divided, when it will be left in connection only with the muscles on its own side of the glottis.

Having thus established the fact of a chiasm between the superior glottic nerves, it was requisite to ascertain whether the inferior or recurrent laryngeal nerves entered into communication with the superior nerves, or whether they possessed any similar interlateral connection of their own.

Experiment.—Snapper, weight six pounds. We divided first the right and left superior laryngeal nerves. The glottis opened as usual, but had lost its power to close firmly.

Section of the right recurrent which followed, as the next step, produced paralysis of the right glottic lip.

Galvanization of one recurrent caused opening of only the corresponding lip of the glottis. Repetitions of the above experiment led to no different result.

Order of section, and results:—

1. Section of both superior laryngeal nerves; loss of closing power.

2. Section of right inferior laryngeal nerve; loss of opening power in right lip of glottis.

We inferred from the above stated experiment and the repetitions of it, that no interlateral nerve fibres connected the two inferior laryngeal nerves. Furthermore, we failed to discover any branch to which such a function could have been assigned.

The object of the very extraordinary and really exceptional arrangements, which we have here pointed out, is not altogether clear. We arrive only at the general conclusion, that the integrity of the glottic function in turtles, appears to have been guarded with unusual care. Why this should be the case in aquatic chelonians it is easy to understand, but the necessity for it in terrestrial species seems to us less obvious, yet it is as perfect in the box turtle as in the emydæ and chelonuræ. Perhaps the need for such precautions in all may be due to the fact that all retain the inspired air during long periods, even when on land. Paralysis of the closing power of the glottis would allow the air to escape instantly, and would oblige the animal to make repeated and therefore laborious inspiratory efforts. Paralysis of the opening power would insure death from apnœa. Hence we have two sets of nerves controlling the opening muscles. One entire set may be destroyed and yet respiration continue. Even one of those remaining, if these be the upper nerves, may be lost, and still the glottis fulfil its entire duty in the train of breathing movements. Thus, also, in regard to the closing power. The elasticity of the glottic lips is one agent, although but a subsidiary one. Then we have the interlateral communication between the two superior laryngeal nerves, which alone can forcibly close the chink of the glottis. By virtue of this true chiasm one of these nerves being injured, the other is ample to effect the normal purpose of both.

Nor is it less curious to observe how artfully the whole apparatus has been guarded against accident.

The lower or recurrent laryngeal nerves lie alongside of the trachea, sheltered by its projecting form. The superior nerves are protected in their course by the superior hyoid cornu, and the larynx and its singular nervous circle are deeply buried beneath, or rather above the strong bony and cartilaginous body of the hyoid bone. Nature seems to have been lavish of expedients for securing the safety of these most important parts.

Before leaving this portion of our subject, it may not be amiss to state that we have made a number of experiments on birds and mammals, to ascertain whether any such chiasm exists in the glottic nerves of these animals. But in all cases section of one motor nerve caused loss of movement in its own side of the larynx, and we therefore conclude that this arrangement does not extend to the classes in question. Whether or not it is to be found in Batrachia and ophidian reptiles, we have not as yet ascertained.

The remaining physiology of the pneumogastric nerve in turtles is not less obscure than in other animals. As in these latter, so in turtles, it sends branches to the trachea, lungs and heart.

We have cut the nerve in a number of turtles, some of whom survived upwards of a mouth and then exhibited no marked evidence of diseased lungs. In others, there was occasionally found an abscess at the base of the neck. This pathological occurrence is, however, a common one in turtles caught with the hook, and cannot, with any probability, be supposed to be due to the section of the pneumogastric.

The only striking effect of this section was, the constant sensibility which the nerve then exhibited. At the moment of dividing or crushing it, the animal showed every possible evidence of acute pain. Irritation of the centric end of the cut nerve gave rise to like phenomena, while stimulation of the peripheral end caused no such results.

A number of careful experiments were made to ascertain whether these irritations of the nerve produced any instant effect, either upon the inspiratory or expiratory muscles of the breast-box. But in no case did the stimulation seem to influence them to movement.

Galvanization of the pneumogastric nerve in turtles arrests the heart's movements. Gentle irritation of the trunk causes the heart to beat more rapidly. Section of one nerve causes the heart to quicken its pulsations. Division of both nerves induces still more rapid action, but in either case the heart, after a few hours, regains its original rate of pulsation.

The nerves which supply motor endowments to the internal respiratory muscles need no special illustration here. They are fully described in the anatomical section of this essay. It only remains to add, that their office and relation to the muscles was tested by stimulating them with galvanism and by dividing them, so as to cause paralysis of the muscles in question.

The centre, to which proceed impressions, giving rise therein to respiratory impulses, appears to be, as in other animals, the medulla oblongata. The site of the respiratory ganglions would scarcely have attracted our attention, however, had it not been, that, in the following experiment, a fact was noticed which induced us to examine the question more fully.

Experiment.—In a turtle previously used to examine the offices of the laryngeal nerves, and in whom the glottis could still open on one side, we divided the cervical spine at its upper third, and continued to watch the respiratory muscles. To our surprise the flank muscles acted at intervals for thirty minutes, but the two sides no longer moved synchronously. At one moment the right muscle contracted, at another the left, and the movements of both were irregular and sometimes incomplete.

It appeared to us, that these motions after section of the spine might be merely the rhythmic repetition of habitual movements, such as, according to Brown-Séquard, appear sometimes in the diaphragms of mammals even. Long after these muscles in the turtle ceased to move, all the other reflex acts continued, and excepting these, almost every muscle below the point of section could be excited easily to reflex motion; neither was there any longer a synchronism of action between the respiratory muscles of the glottis and those of the breast-box.

Experiment.—Turtle, weight six pounds. In this case, also, the cervical spine was divided, but although the reflex activity of most of the parts below the section was remarkable, the respiratory muscles alone failed to respond to excitation of distant parts. During the spasm caused by the section of the spine, the expiratory muscles, contracting, emptied the lungs, which were not again filled with air.

Experiment.—Turtle, weight 4 pounds. The sympathetic nerves on both sides, in this turtle, had been cut several weeks, and the wounds in the neck were nearly healed. The animal seemed well and very active. The cervical spine was divided with little loss of blood. General spasm ensued, the glottis opened, expiration followed, but no after inspiration, and the glottis closed. During an hour no inspiration occurred, although the glottis opened and shut at about the usual respiratory intervals. To make more sure of this, the trachea was cut across, the lung fully inflated, and a tube secured in the lower end of the trachea. Through a short caoutchouc tube the trachea was thus connected with Poiseuille's hæmadynamometer, filled to its 0° with mercury; on turning a stopcock the column rose about two millimetres, the glottis continuing in repose. Then the glottis opened, but no synchronous contraction of the lung muscles took place; indeed, the slightest must have been indicated instantly by the mercurial column. During frequent repetitions of glottic motion, no correspondent activity was at any time exhibited by the respiratory muscles of the breast-box. It follows, therefore, that while the flank respiratory muscles may after separation from their nerve centres move for a time, as do other habitually rhythmical muscles like the heart, that these motions do not occur in all cases, and that they are plainly not dependent on a respiratory centre below the line of spinal section.

The regular movements of the glottis were, as we we have shown, uninterrupted by the section of the cervical spine. The question arose as to the exciting cause of these motions. That they were not due to impulses propagated through the main trunks of the pneumogastric nerves, was shown by their continuance after the successive division of these two nerves below the origin of the glottic nerves. It thus became plain that the medulla must receive its excitations from the head alone, perhaps through the fifth pair of nerves, which acted as afferent trunks, the motor nerves of the larynx completing the nervous circle as efferent branches. Hence the continued action of the glottis after division of the cervical spine.

The principal points in the foregoing paper to which we desire to draw attention as novelties are as follows:—

1st. In Chelonians the superior laryngeal nerve is distributed both to the opening and closing muscles of the glottis.

2d. The inferior laryngeal nerve is distributed solely to the opening muscle of the glottis.

3d. A true chiasm exists between the two superior laryngeal nerves.

4th. The expiratory muscle lies within the breast-box, and consists of anterior and posterior bellies connected by a strong tendon continuous across the middle line, and common to both sides of the animal.

5th. The inspiratory muscles occupy the flank spaces on either side.

6th. Inspiration is effected by the contraction of the flank muscles, which in appearance strongly resemble the diaphragms of superior animals.

7th. Expiration is effected by the consentaneous action of the four muscular bellies above described, which thus compress the viscera against the lungs. The act of respiration consists of an expiration and an inspiration, during which the glottis remains open.

8th. The opening of the glottis is effected through the agency of the superior and inferior laryngeal nerves, both of which are distributed to the dilating muscle of the glottis. The superior laryngeal nerve presides over the closure of the glottis, being in part distributed to its sphincter muscle. The elastic contractility of the glottic cartilages aids in closing this orifice. After section of the superior laryngeal nerves, the glottis may still be opened by the agency of the inferior laryngeal nerves, its imperfect closure being then effected by means of the elasticity of its cartilaginous lips. The chiasm of the superior laryngeal nerves enables one of these nerves to open and shut the glottis after section or disease of the opposite nerve and of both inferior laryngeals.

Physiologists have therefore been in error when describing the respiration of Chelonians as analogous to that of Batrachians, since it far more closely resembles the breathing of the higher vertebrates.

APPENDIX.

Since committing to the press the preceding paper, we have had the opportunity of examining the respiratory apparatus of one of the Trionychidæ.[1] The striking characters of this family and its border rank amongst fresh-water turtles, render a knowledge of its respiratory structure of peculiar interest, and constitute our apology for this appendix.

Amyda mutica, Fitz.—The general plan of arrangement of the respiratory muscles is the same as heretofore described, the inspiratory muscles occupying the flank spaces, and the expiratory muscle being attached to the dorsal shield and inclosing the viscera. The origin of each inspiratory muscle, in detail, is as follows: From the bony edge of the plastron, from the carapace at the line of termination of the ribs, from the fascia lata, where the thigh bounds the flank space, from the spinous process of the ilium and thence to the place of beginning on the plastron. The central tendon into which these fibres are inserted, is a mere raphæ posteriorly, but widens anteriorly into a lance-shaped extremity, one-fourth of an inch wide at its widest part. The whole muscle is relatively large compared with that of other species. The expiratory muscle presents greater variation in origin and form than the inspiratory muscle; the fibres are longer than in other turtles examined, and the anterior and posterior bellies broader. These bellies meet at their outer margins and leave no space under the bridge of the plastron where the muscular fringe is absent when the muscle is viewed from below, as exists in the snapper, Fig. 3. The effect of this widened muscular margin is to diminish the size of the common tendon, and reveal its true character more strikingly than in other families. On dividing the tendon and removing the viscera, the muscular fibres are observed spreading out across the dorsal shield in fan-like radii from each intercostal space, from the first to the sixth inclusive. The third intercostal space gives attachment to the strongest fibres, causing it to appear as the centre from which the whole muscle radiates, the fibres running forward and outward across the area of a quarter circle constituting the anterior belly, and those running backwards over a similar area, the posterior belly. More precisely, however, the origin of the anterior belly is from the first and second intercostal spaces and from the costal margin of the third space;

---

[1] For living specimens of *Amyda mutica*, we are indebted to Mr. Robert L. Walker, of Allegheny County, Pennsylvania.

6

from these points the fibres extend in a prevailing direction forwards, over the anterior quarter shell towards its periphery, and then, arching around the viscera, are inserted into the central tendon. At the line where the muscle leaves the shell it receives additional fibres. The posterior belly arises from the sixth, fifth, and fourth intercostal spaces, and from the vertebral margin of the third space (overlying the fibres of the anterior belly that come from this space), and stretches over the posterior quarter circle of the shell, to be inserted into the common tendon. It receives reinforcements of fibres where it leaves the shell to inclose the viscera, as does the anterior belly.

PUBLISHED BY THE SMITHSONIAN INSTITUTION,

WASHINGTON, D. C.

APRIL, 1863.